Modeling and Simulation of Intelligent Transportation Systems

As transport networks become more congested, there is a growing need to adopt policies that manage demand and make full use of existing assets. Advances in information technology are now such that intelligent transportation systems (ITSs) offer real potential to meet this challenge by monitoring current conditions, predicting what might happen in the future, and providing the means to manage transport proactively and on an area-wide basis. *Modeling and Simulation of Intelligent Transportation Systems* provides engineers, professionals, and researchers an intuitive appreciation for ITS theory, related sensor technologies, and other practical applications, including traffic management, safety, design optimization, and sustainability.

- Provides the theory and practical applications of Intelligent Transport Theory which will be helpful as highway construction recedes as a sustainable long-term solution.
- Includes several case studies that illustrate the concepts presented throughout.

Taylor and Francis Series in Resilience and Sustainability in Civil, Mechanical, Aerospace and Manufacturing Engineering Systems

Series Editor
Mohammad Noori
Cal Poly San Luis Obispo

PUBLISHED TITLES

Automation in Construction Toward Resilience: Robotics, Smart Materials and Intelligent Systems
Edited By Ehsan Noroozinejad Farsangi, Mohammad Noori, Tony T.Y. Yang, Paulo B. Lourenço, Paolo Gardoni, Izuru Takewaki, Eleni Chatzi, and Shaofan Li

Data Driven Methods for Civil Structural Health Monitoring and Resilience: Latest Developments and Applications
Edited By Mohammad Noori, Carlo Rainieri, Marco Domaneschi, Vasilis Sarhosis

Navigating the Complexity Across the Peace-Sustainability-Climate Security Nexus
Bernard Amadei

Seismic Resilience Assessment of Hospital Infrastructure
Qingxue Shang, Tao Wang, Jichao Li, and Mohammad Noori

Modeling and Simulation of Intelligent Transportation Systems
Wael A. Altabey, Mohammad Noori, Ahmed Silik, Marco Domaneschi, and Weixing Hong

Series Editor Bio: Prof. Mohammad Noori is a professor of mechanical engineering at California Polytechnic State University, San Luis Obispo. He received his BS (1977), his MS (1980), and his PhD (1984) from the University of Illinois, Oklahoma State University, and the University of Virginia respectively; all degrees in Civil Engineering with a focus on Applied Mechanics. His research interests are in stochastic mechanics, nonlinear random vibrations, earthquake engineering, and structural health monitoring, AI-based techniques for damage detection, stochastic mechanics, and seismic isolation. He serves as the executive editor, associate editor, technical editor, or a member of the editorial boards of 8 international journals. He has published over 250 refereed papers, has been an invited guest editor of over 20 technical books, has authored/co-authored 6 books, and has presented over 100 keynote and invited presentations. He is a Fellow of ASME and has received the Japan Society for Promotion of Science Fellowship.

For more information about this series, please visit: https://www.routledge.com/ Resilience-and-Sustainability-in-Civil-Mechanical-Aerospace-and-Manufacturing/ book-series/ENG

Modeling and Simulation of Intelligent Transportation Systems

Wael A. Altabey, Mohammad Noori, Ahmed Silik,
Marco Domaneschi, and Weixing Hong

CRC Press
Taylor & Francis Group
Boca Raton London New York

CRC Press is an imprint of the
Taylor & Francis Group, an **Informa** business

Designed cover image: Shutterstock

First edition published in 2025
by CRC Press
2385 NW Executive Center Drive, Suite 320, Boca Raton FL 33431

and by CRC Press
4 Park Square, Milton Park, Abingdon, Oxon, OX14 4RN

CRC Press is an imprint of Taylor & Francis Group, LLC

© 2025 Wael A. Altabey, Mohammad Noori, Ahmed Silik, Marco Domaneschi, and Weixing Hong

ISBN: 978-1-032-69174-9 (hbk)
ISBN: 978-1-032-69180-0 (pbk)
ISBN: 978-1-032-69178-7 (ebk)

DOI: 10.1201/9781032691787

Typeset in Times
by codeMantra

Contents

Preface

A safe and convenient travel environment is the people's good expectation for the transportation industry. In order to comprehensively improve the safety level of transportation facilities, an intelligent transportation system (ITS) proposes to address the problems of mobility, environmental pollution, and road safety, as well as their related applications. ITS is a sophisticated application that aims to provide innovative services related to various modes of transportation and traffic management to better inform users and enable them to make safer, more coordinated, and agile use of transportation networks. As a result, ITSs are being created worldwide to reduce these losses by implementing smart technologies such as traffic management, ramp meters, and traveler information systems, among other things. Furthermore, with the concept of smart cities changing cities into digital societies and making citizens' lives simpler in every manner, the ITS has become an integral part of all.

This book presents intelligent solutions to transportation problems using many advanced technologies, such as communication, sensing, and control, which are used for managing a large amount of information. The applications vary and include fleet management, driving behavior, traffic control, trajectory planning, connected vehicles, and energy consumption efficiency. This book provides a comprehensive application and guidelines for ITSs. It is a good educational resource for engineering students at both undergraduate and graduate levels, and for practicing engineers. The ITSs analysis is a multidisciplinary subject. The scope is not limited to any particular engineering discipline. This book also intends to provide engineers and researchers with an intuitive appreciation for ITSs theory, intelligent transport sensors, and intelligent transport methods. The application of ITSs theory in transport engineering is relatively insufficient. In practice, the ITSs theory can be utilized for the safety evaluation of a bridge, traffic control, and optimization design of the bridge.

Authors' Biography

Wael A. Altabey, Mohammad Noori, Ahmed Silik, Marco Domaneschi, Weixing Hong

Wael A. Altabey is a research associate professor since 2018 at the International Institute for Urban Systems Engineering (IIUSE), Southeast University, Nanjing, China, and the National and Local Joint Engineering Research Center for Basalt Fiber Production and Application Technology, Southeast University, Nanjing, Jiangsu, China, after completing a postdoctoral research fellowship for 2 years (2016–2018). Before that he was an associate professor at the Department of Mechanical Engineering, Alexandria University, Alexandria, Egypt. Since 2016 his research has focused on the utilization of artificial intelligence (AI)-based schemes for structural health monitoring (SHM) and non-destructive testing (NDT) for damage classification, detection, diagnosis, prediction, and dynamic response analysis in composite and steel structures (such as aircraft, wind turbines, pipes, bridges, and industrial machines) at National and Local Joint Engineering Research Center for Basalt Fiber Production and Application Technology, Southeast University, Nanjing, Jiangsu, China. This is the only national R&D platform awarded by the National Development and Reform Commission in this industry with more than 30 nationally authorized patents. The center's international and national awards indicators have reached international and local leading levels, filling many technical gaps in China. He participated in several research activities, which were achieved from NSFC and private sectors. He was listed in the Stanford List of World's Top 2% Scientists from 2020, until now. He is a member of the Intelligent and Resilient Infrastructure (IRI) Center, which is the official center affiliated with the Southeast-Leeds-Cambridge universities collaboration program. He was a research scientist at Nanjing Zhixing Information Technology Co., Ltd., Nanjing, Jiangsu, China. He serves on various technical committees in several international conferences and workshops, as the guest editor of special issues in several international scientific journals, and serves on the editorial board of several international scientific journals in the field of artificial intelligence, mechanical, materials, and civil engineering. He is a peer reviewer of more than 200 international scientific journals. He is an author and a co-author of more than 110 high-impact journal papers, 60 scientific conference papers and 40 chapters, 10 academic and research books, and patents, and delivered over 40 invited talks. His research interests include Smart and Nanomaterials, Composite Structures, Structural Health Monitoring (SHM), Artificial Intelligence (AI), Non-Destructive Testing (NDT), Digital Twins Model of Structural Behavior, System Identification, Damage Detection: Vibration-Based Techniques, Fiber Optical Sensing Technique, Structural Control, Structural Resilience and Reliability, Hysteretic Systems, Micro/Nano Electro Mechanical Systems (MEMS/NEMS), and Energy Harvesting Model for Self-Powered Sensors.

Mohammad Noori is a professor of Mechanical Engineering at Cal Poly, San Luis Obispo, a fellow and life member of the American Society of Mechanical Engineering, and a recipient of the Japan Society for Promotion of Science Fellowship. His work in nonlinear random vibrations, especially hysteretic systems, in seismic isolation and application of artificial intelligence methods for structural health monitoring, is widely cited. He has authored over 300 refereed papers, including over 150 journal articles; has published 15 scientific books and 31 book chapters in archival volumes; has edited 15 technical books; and has been the guest editor of 15 journal volumes and proceedings. He was a co-founder of the National Institute of Aerospace, established through a $379 million 15-year NASA contract in partnership with NASA Langley Research Center. He has also received over $14 million in support of his research from NSF, ONR, National Sea Grant, and industry. He has supervised 24 postdoctoral, 26 PhD, and 53 MS projects. He has given over 20 keynote and 76 invited talks and lectures. He is the founding executive editor of a scientific journal; serves on the editorial board, or as the associate editor, of over 15 other journals; and has been a member of the scientific committee of numerous conferences. He directed the Sensors Program at the National Science Foundation in 2014, has been a distinguished visiting professor at several highly ranked universities in Europe and Asia, and serves as the scientific advisor for several organizations and technical firms. He was the dean of engineering at Cal Poly and served as a chaired professor and department head at NC State University and WPI and as the chair of the National Committee of Mechanical Engineering Department Heads. He has developed a unique online course, How to Write an Effective Research Paper, offered by Udemy.com, taken by over 9,000 students worldwide. He is an elected member of Sigma Xi, Pi Tau Sigma, Chi-Epsilon, and Sigma Mu Epsilon honorary research societies. In 1996, Noori was invited by President Clinton's Special Commission on Critical Infrastructure Protection and presented a testimony as a national expert on that topic. Noori is the Founding Editor of Resilience and Sustainability in Civil, Mechanical, Aerospace and Manufacturing Engineering Systems Series of CRC Press/Taylor and Francis.

Ahmed Silik is a recent doctoral recipient from Southeast University, Nanjing, China, and currently works as a research associate at IIUSE as part of Professor Zhishen Wu's group. Before that, he was a lecturer at the Department of Civil Engineering, Nyala University, Nyala, Sudan, and worked as a consultant civil engineer in Sudan. He is a fellow of the Sudanese Engineering Council. His main research interests are focused on the utilization of wavelet transform and AI-based techniques for structural health monitoring, damage identification, and big data analysis in civil structural engineering. He has been actively involved in several research topics and activities in the area of SHM related to damage prediction and assessment of structures and bridges, including a collaborative project with colleagues at Saitama University, Japan. He has also been involved in an NSFC project on dynamic response assessment and modeling of hysteretic systems. He has published several peer-reviewed journal manuscripts and attended several scientific conferences.

Marco Domaneschi is currently an assistant professor at the Department of Structural, Geotechnical, and Building Engineering of Politecnico di Torino, where he teaches courses on smart infrastructures, earthquake engineering, and structural design. Formerly, he was a research associate and appointed professor of structural engineering at Politecnico di Milano. He is a professional structural engineer for special structures and serves as an R&D consultant in industrial manufacturing and mechanical engineering. He received his PhD from the University of Pavia (2006) and was a visiting researcher at several global universities (US, China, E). He currently serves as an associate editor and editorial board member for several international journals such as the *Journal of Vibration and Control, Bridge Engineering ICE-UK, Advances in Engineering Software, Journal of Traffic* and *Transportation Engineering*. He is also a member of several research associations such as the International Society of Structural Health Monitoring of Intelligent Infrastructure (ISHMII) and the International Association for Bridge Maintenance and Safety (IABMAS). He is also a reviewer for more than 40 international journals. He has been a speaker, sessions chair, editorial board member, and organizer at several international conferences. He received numerous awards for best presentations at conferences, and research papers and activities. He supports/supported the coordination of several research projects and has/had scientific responsibility in numerous research projects. He has authored over 70 journal articles and 130 international conference papers. His research interests and activities include structural control and health monitoring, resilience, sustainability and robustness of structures and communities, earthquake engineering and seismic risk, special structures, small- and large-scale simulations, emergency evacuation, and structural collapse analysis. He is included from 2020 in Stanford's list of the top 2% most-cited scientists published by Elsevier and a member of the Teaching Board of the National Doctoral Programs.

Weixing Hong serves as the Chairman of Nanjing Zhixing Information Technology Co., Ltd., the Executive Dean of Jiangsu Advanced Transportation Institute, and a Senior Engineer. He also serves as the Chairman of the Enterprise Committee of the ISHMII International Structural Health Monitoring Association, a member of the ISO TC204 World Intelligent Transportation International Standard Expert Group, a member of the Autonomous Driving Working Committee of the China Highway Society, and a member of the Information Technology Working Committee of the Jiangsu Comprehensive Transportation Society. Nanjing Zhixing Information Technology Co., Ltd. and Jiangsu Advanced Transportation Institute are high-tech enterprises that integrate industry, academia, and research. With the core technologies of the IoT, Cloud Computing, Big Data, and AI, and relying on the industry background of traffic analysis and management, they are committed to combining 'Intelligent+' technology with transportation infrastructure construction and maintenance, providing equipment and platform services centered on transportation intelligent sensing and industry application algorithms.

1 Intelligent Transport Systems

1.1 BACKGROUND

As transport networks become more congested, and new highway construction recedes as a sustainable long-term solution, there is a growing need to adopt policies that manage demand and make full use of existing assets.

Advances in information technology are now such that intelligent transport system (ITS) offers real possibilities for authorities to meet this challenge by monitoring what is going on, predicting what might happen in the future, and providing the means to manage transport proactively and on an area-wide basis.

This chapter focuses on ITS measures. First, a description of ITS objectives (benefits) is provided, followed by a general description of ITS measures. Thereafter, an explanation is given as to why this thesis focuses on infrastructure-related systems, in particular highways. The benefits of highway ITSs, based on estimates via modeling studies, as well as measurements of before and after studies, are described.

The information presented in this chapter aims to answer the following questions:

- Which benefits of ITS have been established in the developed world?
- Which ITS systems are potentially beneficial for your country?

1.2 OBJECTIVES OF ITS MEASURES

Importantly, ITS can facilitate the delivery of a wide range of policy objectives, beyond those directly associated with transport, bringing significant benefits to transport users and those who live and work within the area. There are six main objectives/benefits that have been identified in the international literature.

Safety: An explicit objective of the transportation system is to provide a safe environment for travel while continuing to strive to improve the performance of the system. Although undesirable, crashes and fatalities are inevitable occurrences. Several ITS services aim to minimize the risk of crash occurrence. This objective focuses on reducing the number of crashes and lessening the probability of a fatality should a crash occur. Typical measures of effectiveness used to quantify safety performance include the overall crash rate, fatality crash rate, and injury crash rate. ITS services should also strive to reduce the crash rate of a facility or system. Crash rates are typically calculated in terms of crashes per year, crashes per million vehicle kilometers traveled, or crashes per 10,000 inhabitants.

Mobility: Improving mobility (and reliability) by reducing delay and travel time is a major objective of many ITS components. Delay can be measured in

DOI: 10.1201/9781032691787-1

many different ways, depending on the type of transportation system being ana-lyzed. Delay of a system is typically measured in seconds or minutes of delay per vehicle. Also, delays for users of the system may be measured in person hours. Delay for freight shipments could be measured in time past the scheduled arrival time of the shipment. Delay can also be measured by observing the number of stops experienced by drivers before and after a project is deployed or imple-mented. Travel time variability indicates the variability in overall travel time from an origin to a destination in the system, including any modal transfers or en-route stops. This measure of effectiveness can readily be applied to inter-modal freight (goods) movement as well as personal travel. Reducing the variability of travel time improves the reliability of arrival time estimates that travelers or companies use to make planning and scheduling decisions. By improving operations and incident response, and providing information on delays, ITS services can reduce the variability of travel time in transportation networks. For example, traveler information products can be used in trip planning to help re-route commercial drivers around congested areas resulting in less variability in travel time.

Efficiency: Many ITS components seek to optimize the efficiency of existing facilities and use of rights-of-way so that mobility and commerce needs can be met while reducing the need to construct or expand facilities. This is accomplished by increasing the effective capacity of the transportation system. Effective capacity is the "maximum potential rate at which persons or vehicles may traverse a link, node or network under a representative composite of roadway conditions," including "weather, incidents and variation in traffic demand patterns". Capacity, as defined by the Highway Capacity Manual, is the "maximum hourly rate at which persons or vehicles can reasonably be expected to traverse a given point or uniform section of a lane or roadway during a given time period under prevailing roadway, traffic, and control conditions". The major difference between effective capacity and capacity is that capacity is generally measured under typical conditions for the facility, such as good weather and pavement conditions, with no incidents affecting the system, while effective capacity can vary depending upon these conditions and the use of manage-ment and operational strategies.

Throughput is defined as the number of persons, goods, or vehicles traversing a roadway section or network per unit time. Increases in throughput are sometimes realizations of increases in effective capacity. Under certain conditions, it may reflect the maximum number of travelers that can be accommodated by a transportation sys-tem. Throughput is more easily measured than effective capacity and, therefore, can be used as a surrogate measure when analyzing the performance of an ITS project. The reader needs to bear in mind that local circumstances influence local capacities, as well as measured throughputs.

Productivity: ITS implementation frequently reduces operating costs and allows productivity improvements. In addition, ITS alternatives may have lower acquisition and life cycle costs compared to traditional transportation improve-ment techniques. The measure of effectiveness for this objective is cost sav-ings as a result of implementing ITS. Another way to view the cost savings is to quantify the cost savings between traditional and ITS solutions to address-ing problems. Energy and environment: The air quality and energy impacts of

ITS services are very important considerations, particularly for non-attainment areas. In most cases, environmental benefits can only be estimated by the use of analysis and simulation. The problems related to regional measurement include the small impact of individual projects and large numbers of exogenous variables including weather, contributions from non-mobile sources or other regions, and the time-evolving nature of ozone pollution. Small-scale studies generally show positive impacts on the environment, and these impacts result from smoother and more efficient flows in the transportation system. However, the environmental impacts of travelers reacting to large-scale deployment in the long term are not well understood. Decreases in emission levels and energy consumption have been identified as measures of effectiveness for this objective. Customer satisfaction: Given that many ITS projects and programs were specifically developed to serve the public, it is important to ensure that user (i.e., customer) expectations are being met or surpassed. Customer satisfaction measures and characterizes the distance between users' expectations and experiences in relation to a service or product. The central question in a customer satisfaction evaluation is, "Does the product deliver sufficient value (or benefits) in exchange for the customer's investment, whether the investment is measured in money or time?" Typical results reported in evaluating the impact of customer satisfaction with a product or service include product awareness, expectations of product benefit(s), product use, response (decision making or behavior change), realization of benefits, and assessment of value. Although satisfaction is difficult to measure directly, measures related to satisfaction can be observed including the amount of travel in various modes, and the quality of service, as well as the volume of complaints and/or compliments received by the service provider.

In addition to user or customer satisfaction, it is necessary to evaluate the satisfaction of the transportation system provider or manager. For example, many ITS projects are implemented to improve coordination between various stakeholders in the transportation arena. In such projects, it is important to measure the satisfaction of the transportation provider to ensure the best use of limited funding. One way to measure the performance of such a project is to survey transportation providers before and after a project was implemented to see if coordination was improved. It may also be possible to bring together providers from each of the stakeholder groups to evaluate their satisfaction with the system before and after the implementation of an ITS project.

1.3 DESCRIPTION OF ITS MEASURES

ITS is a very broad field. It varies from traffic light control to incident management, from enforcement to passenger information, and driver assistance to intelligent speed limit enforcement. Structuring a broad field like ITS measures is difficult as each structure can be challenged. We can divide ITS measures into three groups:

1. Intelligent Traffic Management Systems measure and analyze traffic flow information and take ITS measures to reduce problems. They consist of computerized traffic signal control, highway, and traffic flow management

systems, electronic licensing, incident management systems, electronic toll and pricing, traffic enforcement systems, and intelligent speed adaptation.

2. Intelligent Passenger Information Systems improve the knowledge base of Customers and consist of passenger information systems, in-vehicle route guidance systems, parking availability guidance systems, digital map databases, and variable messaging systems.

3. Intelligent Public Transport Systems include ITS measures that aim to improve public transport performance. They consist of intelligent vehicles, Intelligent Speed Adaptation, transit fleet management systems, transit passenger information systems, electronic payment systems, electronic licensing, transportation demand management systems, and public transport priority.

As mentioned in Section 1.2, safety, mobility, efficiency, productivity, energy and environment, and customer satisfaction are the benefits internationally identified for ITS measures. The focus of this book is on transportation benefits. It was, therefore, decided to exclude productivity, as it aims to produce economic benefits.

Energy and environmental benefits generally focus on benefits with regard to natural resources and are secondary effects. This study focuses primarily on the impacts of measures. The characterization of benefits in this book is as follows:

• Safety: Safety-related ITS measures aim to reduce accidents and dangerous situations,
• Mobility and efficiency: These measures aim to optimize the use of road capacity, reduce unnecessary and inefficient driving, and
• Customer satisfaction: the provision of information, security, etc.

Table 1.1 provides an overview of the ITS-identified measures.

TABLE 1.1
Overview of ITS Measures per Objective

	Intelligent Traffic Management Systems	Intelligent Passenger Information Systems	Intelligent Public Transport Systems
Safety	• Variable speed limits • Lane management • Incident management • Warning systems • CCTV cameras • Automatic vehicle • identification • Intelligent Speed • Adaptation • Weight in motion	• Navigation systems • Parking guidance • Cruise control • Warning systems • Intelligent Speed • Adaptation • Black-box systems • Automated vehicle • identification • Docking systems • Distance warning	• Fleet management • Navigation systems • Electronic ticketing • CCTV cameras • High-speed ground • transportation • Automatic vehicle • identification • Intelligent Speed • Adaptation • Distance warning

(Continued)

TABLE 1.1 (Continued)
Overview of ITS Measures per Objective

	Intelligent Traffic Management Systems	Intelligent Passenger Information Systems	Intelligent Public Transport Systems
Mobility and efficiency	• Variable speed limits • Lane management • Incident management • Warning systems • CCTV cameras • Ramp metering • Traffic control • Electronic toll • Collection • Real-time information • Parking guidance	• Navigation systems • Parking guidance • Cruise control • Warning systems	• Public transport • priority • Fleet management • Navigation systems • Electronic ticketing • System integration • High-speed ground • transportation • Real-time • Information
Customer satisfaction	• CCTV cameras • Lane management • Warning systems • Electronic toll • collection • Real-time information • Parking guidance	• Navigation systems • Parking guidance • Real-time information • Electronic toll • collection • Docking systems • Warning systems	• Real-time information • System integration • Electronic ticketing • CCTV cameras

1.3.1 INTELLIGENT TRAFFIC MANAGEMENT SYSTEMS

Government and road agencies are responsible for the provision of infrastructure and infrastructure-related (ITS) systems enhancing road safety, mobility, etc. Systems for highways and secondary roads are generally different. Examples of infrastructure-related ITS systems are:

- Variable Speed Limits (VSL): Variable Message Signs (VMS) are mostly used to apply VSL on a road. The aim is to reduce the speed before congestion appears, which will result in a more homogenized traffic flow (efficiency),
- Lane management: dedicated lanes for trucks, buses, and High-Occupancy Vehicles (HOV) are commonly used in the developed world to improve the road system,
- Incident management: Incidents have a negative impact on the traffic flow handling of a road. Better incident handling procedures can limit the time factor of an incident. Moreover, predicting the risk of an incident will also help to clear the incident quicker,
- Warning systems: These systems are able to provide several types of information (like fog, congestion, incidents, etc.). There are several communication systems (i.e. beacon-vehicle or GPS-based systems) that can be used for the development of warning systems,
- CCTV cameras: Closed Circuit TeleVision (CCTV) cameras take video or photo shots of previously identified situations. The general aim is to

"smoothen" traffic flows. Virtual loops are often used to analyze video material automatically,

- Automated Vehicle Identification (AVI): The general trend is that automated vehicle identification is done using tags, which can help law enforcement,
- Intelligent Speed Adaptation (ISA): ISA is a collective name for systems in which the speed of a vehicle is permanently monitored within a certain area. When the vehicle exceeds the speed limit, the speed is automatically adjusted or a warning is provided to the driver,
- Weight In Motion (WIM): These systems are currently used in your country, whereby technology assists in checking trucks with regards to overloading,
- Ramp metering: Ramp metering is a method to limit/regulate the entering of vehicles onto an arterial or highway. Loops in the road are used to measure the flows, and traffic lights let single vehicles through,
- Traffic control: Traffic controllers are used to regulate traffic flows at intersections (see also public transport priority),
- Electronic Toll Collection (ETC): From an infrastructure point of view, the aim is to collect fees in an undisruptive way. Pay lanes and road pricing should ideally be realized using ETC,
- Real-time information systems: Real-time information systems use data collected by traffic management centers to inform road users of incidents, delays, etc., and
- Parking guidance: Parking guidance systems, based on navigation systems, provide drivers with information regarding the availability of parking bays.

1.3.2 INTELLIGENT PASSENGER INFORMATION SYSTEMS

The international car industry has been adding several ITS systems to private vehicles. These systems focus on safety, enforcement and control, mobility and efficiency, pre- and on-trip information, as well as assist in the realization of ticketing and pricing systems. It is not attempted to have a complete list of systems, as new systems are developed and added to the market all the time. Nevertheless, Table 1.1 provides a broad range of systems on the market. Systems included in private and freight vehicles are:

- Navigation systems: Electronic systems that provide road information to the (co)driver,
- Cruise Control (CC): Manual systems make sure that the speed limit is not exceeded. Moreover, more intelligent systems are Adaptive Cruise Control (ACC),
- Warning systems: Warning systems related to the vehicle include anti-collision systems (using sensors), weather systems (via radio, navigation systems, etc.), congestion warnings, fuel efficiency (the so-called econom-eter), speed trap warnings (via LCD or sound), etc. Some of these devices, like congestion and weather warning systems, need real-time information,
- Black-box systems: These systems have been used in aviation for many years, but in the road environment the idea is new. The aim is to analyze the vehicle status and warn the driver if problems might occur. A possible

application is analyzing the status of the truck to be able to warn drivers if the system analysis behavior indicates that the driver is falling asleep,

- Automated Vehicle Identification (AVI): The general trend is that automated vehicle identification is done using tags. The identification feature of a tag can help law enforcement,
- Docking systems: Sensors are used to measure the distance of a vehicle to other vehicles or objects. Using docking systems, users are able to park their vehicles more accurately. Moreover, generally, it is possible to park in smaller spaces (parking bay), and
- Distance warning: Similar to docking systems, sensors are used to warn drivers of vehicles that they are getting too close to other vehicles. This happens in a longitudinal as well as lateral manner. Warning systems take the distance and vehicle speed into account.

Other Intelligent Passenger Information Systems are Parking guidance systems, ISA, real-time information systems, ETC (including road pricing).

1.3.3 INTELLIGENT PUBLIC TRANSPORT SYSTEMS

Several systems have been developed and/or are under (further) development to enhance public transport. These systems include:

- Fleet management: These systems are based on navigation technology, and feedback link with operators is added. The operator will be able to follow the vehicle, analyze driver behavior, and take steps if the behavior is unsatisfactory,
- Electronic ticketing: Electronic ticketing will improve the efficiency of a public transport system (payments are made quicker) and will provide a safer environment (the user ID is available and it will be more difficult for criminals to stay anonymous). Moreover, from a traffic engineering point of view, electronic ticketing provides opportunities to improve the collection of travel demand data,
- High-speed ground transportation: These are guided systems that are capable of sustaining operating speeds in excess of 200 km/h. High-speed trains were first used in Japan about 30 years ago. Due to their success in Japan, many European countries started investigating high-speed trains as a viable option,
- Public Transport Priority (PTP): These systems minimize the negative impact of traffic lights for public transport. Many traffic control systems, like the Split, Cycle and Offset Optimizer Technique (SCOOT) system, are able to give PTP at intersections. It is, however, necessary that public transport vehicles can be identified (for example via tags),
- Real-time information: With regards to public transport, real-time information can be used as a Travel Demand Management (TDM) tool. TDM is finding ways to influence human behavior and encourage a shift from private to public transport. ITS systems assist to generate this shift, and

- System integration: To enhance the attractiveness of public transport versus the private car (i.e. travel time, waiting time at stops etc.) has to be improved. Car drivers get information about public transport via VMS. PTP will help to improve the travel time. System integration will reduce the waiting times and possibly the number of transfers providing dynamic stop information (using VMS).

Other Intelligent Public Transport Systems include: navigation systems, CCTV cameras in vehicles and transit points, AVI, ISA, distance warning and real-time information.

1.4 HOW BENEFICIAL ARE ITS SYSTEMS?

1.4.1 Definition of the Research Framework

Obviously it is not possible to investigate all ITS systems within this book. Research with regards to Intelligent Passenger Information Systems is "controlled" by the car manufacturing industry. Several systems, such as navigation systems, vehicle tracking systems (AVI) and warning systems (including econometers), are commercially available in your country. If systems are financially viable, your country car industry will implement them. It was, therefore, decided to exclude Intelligent Passenger Information Systems from this book.

The majority of countries are dependent on the public transport system, but the quality of this system is mostly poor. It would, therefore, be worthwhile to investigate the benefits of Intelligent Public Transport Systems.

1.4.2 Estimated International ITS Benefits

The main focus of this section is to provide an overview of the benefits of ITS based on ex ante modeling exercises. Studies are mainly European based, although some are from the US.

To provide an impression of the efficiency of ITS measures, data from different studies has been collected and compared. Figure 1.1 summarizes the comparison. The actual information is included in Table 1.2.

Readers must keep in mind that the results of these studies have been obtained using different dynamic models. Moreover, some studies include an increased demand in the future, while others do not. Additionally, the research period of the different studies varies, which influences the findings. Despite these differences, a general impression of the impact of ITS measures can be achieved.

In some cases, vehicle related ITS measures have been included in this overview. The criterion for inclusion was that there must be infrastructure related benefits (i.e. throughput).

In a study carried out by Ludmann et al., 100% of vehicles were equipped with ACC. Estimates, using the PELOPS model, were done for highways as well as urban traffic. Two different systems were tested (the second one reacted slightly smoother). The changes in speed and throughput were calculated. The average speed in the first

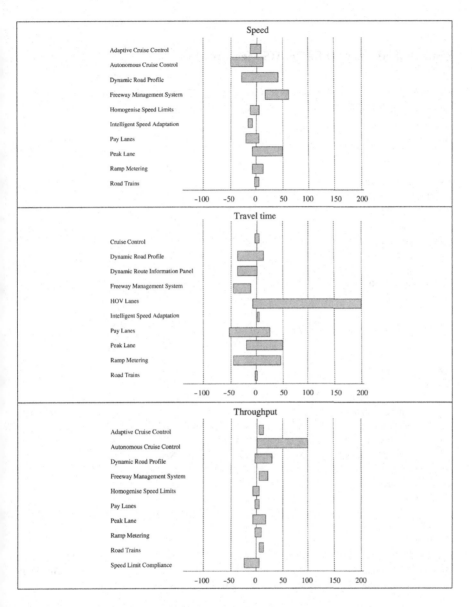

FIGURE 1.1 Estimated international ITS benefits (%).

scenario decreased by 13%, while the speed increased by 6% in the second scenario. Both scenarios showed an increase in throughput (12%–14%).

The effect of Autonomous Adaptive Cruise Control 3 on speed, travel time and throughput (traffic flow) is minor, in a study by Vanderschuren et al. The estimates were obtained for a highway link, using the microscopic simulation model MIXIC. Despite the minor effects on speed, travel time and throughput, a phenomenal reduction in shock waves was estimated, which is an indication of an improvement of the road

TABLE 1.2

Simulated Effects of Different ITS Measures

Measure	Speed	Travel Time	Throughput	Shock Waves[a]	Software
Adaptive Cruise Control Autonomous Adaptive Cruise Control	−13% to +6%	N/A	+12% to +14%	N/A	N/A
• 50%	0%	−1.5%	0%	−80%	MIXIC
• 50% and special lane (SL)	−3%	+1%	+1%	−73%	
• 50% SL and short headways	−1%	0%	+1%	−71%	
• 60% SL and short headways	−1%	−1.5%	−1.5%	−84%	
Autonomous Adaptive Cruise Control					
• 40%	N/A	N/A	+7%	N/A	N/A
• 100%	N/A	N/A	+2%	N/A	
Autonomous Adaptive Cruise Control					
• 100% and 1.5 seconds	N/A	N/A	N/A	N/A	FLOWSIM
• 20% and 1.5 seconds	N/A	N/A	N/A	N/A	
• 40% and 1.5 seconds	N/A	N/A	N/A	N/A	
• 70% and 1.5 seconds	N/A	N/A	N/A	N/A	
Cooperative Adaptive Cruise Control					
• 40%	N/A	N/A	+18%	N/A	N/A
• 100%	N/A	N/A	+100%	N/A	
Dynamic Road Profile	−30%	N/A	30%	N/A	MIXIC
Dynamic Road Profile	−5.9% to	N/A	−1.6% to 17.7%	−4%	INTEGRATION
Dynamic Road Profile	+4.9% N/A	−41% to +16%	N/A	N/A	INTEGRATION
Dynamic Route Information Panel	N/A	−42% to +0%[b]	N/A	N/A	PARAMICS
Freeway Management System	+6 to +62	−48% to −13%	+8% to +25%	N/A	N/A
High-Occupancy Vehicle lane	N/A	−8%	N/A	N/A	N/A
High-Occupancy Vehicle lane	N/A	−1% to +200%	N/A	N/A	N/A

(Continued)

TABLE 1.2 (Continued)
Simulated Effects of Different ITS Measures

Measure	Speed	Travel Time	Throughput	Shock Waves[a]	Software
Homogenize via Speed limits (VMS)	−9.8 to +1.5	N/A	−6.6% to +1.6%	+4.8%	INTEGRATION
Intelligent Speed Adaptation	−10%	+1%	N/A	N/A	D RACULA
Pay Lanes	−15.7% to +3.1%	−56% to −42%	−1.9% to 2.4%	N/A	INTEGRATION
		+15.2 to	0%	N/A	INTEGRATION
• Paying drivers	N/A	+25.3%			
• Non-paying drivers		−33%			
Pay Lanes					
Peak Lane	−5.6% to	N/A	−1.5% to	−4%	INTEGRATION
Peak Lane	+50.7%	−21%	+18.1%	N/A	INTEGRATION
Peak Lane	N/A	−9% to +50%	N/A	N/A	INTEGRATION
	N/A		−5% to +6%		
Ramp Metering	−5.2% to	N/A	−1% to 0.8%	+0.6%	INTEGRATION
Ramp Metering	+8.2%	−21% to			
Morning peak	N/A	+45%[d]	N/A	N/A	INTEGRATION
• Evening peak	N/A	−23% to +22%	N/A	N/A	
Ramp Metering	N/A	−6%	N/A	N/A	INTEGRATION
Ramp Metering	N/A	−48% to −14%	+8%	N/A	INTEGRATION
Road Trains	0%	0%	More freight	0%	MIXIC
Road Trains	N/A	N/A	+10% to +14%	N/A	N/A
Speed limit compliance (from 80% to 100%) Peak					
• Current speed limit	N/A	N/A	+5.6%	N/A	DRACULA
• Speed limit minus 10 kmyh	N/A	N/A	+5.7%	N/A	
Off peak					
• Current speed limit	N/A	N/A	+2.9%	N/A	
• Speed limit minus 10 kmyh	N/A	N/A	−24.2%	N/A	

N/A, not available.

[a] Number of shock waves is indication for the road safety situation; some studies use number of stops.
[b] Estimated; based on different graphs.
[c] The speed is reduced from 120 to 90 km/h.
[d] There is a positive impact on the travel time on the highway and a negative impact on the secondary road network.

safety situation. Another study conducted by VanderWerf et al., estimating the effects for a highway with on and off ramps, indicates that more advanced ACC (Cooperative Adaptive Cruise Control) will have a major impact on throughput. Twice as many vehicles will be able to use the road if 100% of the vehicles have Cooperative ACC.

The study of Marsden et al. indicates that there is an optimal percentage of penetration with ACC. In his modeling exercise for highways, he concludes that optimal penetration is between 10% and 20%. If 40% or more of the vehicles have ACC, the average speed decreases.

Dynamic Road Profiles are applicable to highways. Tampère found a substantial increase in the capacity of the road (indicated by the throughput). In this study, three traditional highway lanes were replaced by four smaller lanes with lower maximum speed during peak hour. An estimated capacity increase of 30% is very promising. The study done by Stemerding indicated that the overall throughput increases by about 5%. In this study, the maximum speed decreased from 100 to 70 km/h. The decrease (4%) in the number of stops is an indication that the road safety situation has improved. Goudappel Coffeng investigated different types of dynamic road profiles. In general, a decrease in travel time was measured (up to 41%). In one case, an increase in travel time of 16% was measured.

Dynamic Route Information Panels are VMS, which inform the driver about congestion ahead, mostly on the highway and/or expected travel times. The study by Van Straaten shows that these types of VMS reduce the severity of congestion. Average travel times decrease by up to 42%. This result is significant. Although it is unknown how many drivers follow the suggested, less congested route, this study indicates that the percentage is high enough to make a difference.

The effects of a freeway management system are very promising. Estimated decreases in travel times up to 48% are remarkable. An estimated increase in the capacity of the highway (throughput) of up to 25% is a striking indicator as well. The results partly appear so positive because a Freeway Management System is a combination of measures:

- Variable Message Signs (VMS);
- Advanced mobile information systems, such as in-vehicle monitoring;
- Automatic toll collection or electronic fare payment;
- CCTV security surveillance (incident management) and vehicle identification, and
- Radio reports, aerial patrols and such.

It would have been interesting if the author had split the effects of the mentioned submeasures.

Comparing this package of measures with the Dynamic Route Information Panel, the difference is minimal. The question is whether the extra effort and costs to provide this package of measures is, therefore, justified.

The estimated effect of HOV lanes created on the highway, concentrate on a change in travel time. A travel time reduction of up to 8% is estimated. Nevertheless, in many instances the travel time, especially for non-HOV vehicles, increases. In the study conducted by Johnston this increase is up to 200%.

Homogenizing via Speed Limits does not always result in more homogenized traffic on a highway system, at least not with the modeled maximum speed limit of 90 km/h. The total throughput in this study decreases by 2%; more traffic is using the secondary road network. Moreover, the number of stops increases, which is negative from a road safety aspect. Analyzing the details, the authors are of the opinion that the limits of the software used might also have influenced the results.

The main aim of ISA is an improvement in road safety. As it is not possible for drivers to exceed the speed limit in the mandatory system modeled, the average speed of vehicles decreases (34% of the vehicles exceeded the speed limit before the introduction of ISA). The modeling was carried out for a network of highways and a limited number of secondary roads. The changes in total travel time are minor. Although this study does not provide shockwave information, it is expected that ISA will reduce shockwaves and, therefore, improve the safety situation on the roads.

Pay lanes are dedicated lanes on highways where a toll is collected. The collection process mostly happens electronically. The estimated effects of pay lanes are generally positive. Schoemakers et al. find an overall decrease in travel time of 33%. Stemmerding et al. find that the decrease is generally for the paying drivers. The non-paying drivers experience an increase in travel time. The effect on the throughput in the study of Vander Werf varies. Both studies focus on highways.

Peak lanes are lanes on the medium of a highway that are only open for traffic during peak hours (could be tidal flow lanes). They are often used in different directions during the morning and evening peak. Peak lanes manage to decrease the congestion risk and keep the flow more homogenized. The total throughput increases by about 5% during peak hour (vehicles currently traveling at other times).

The road safety situation improves slightly. Westra and Bosch also investigated the effects of peak lanes. They found an overall decrease in travel time on the whole network of about 21%. Nevertheless, on parts of the network an increase (+40%) in travel time was found. Bosch et al. found a throughput that varies from –5% to +6%. This study identifies a high risk of increased travel times (between –9% and +50%).

Ramp metering is the application of traffic controllers on on-ramps to reduce disturbance and shockwaves on a highway. First implemented in Chicago, Detroit and Los Angeles, they have been deployed in at least 20 areas in the United Stated. Efficient use of ramp metering can reduce total system travel time, accidents, fuel consumption and vehicular emissions.

Many studies determined that proper ramp metering results in a better overall traffic flow during periods of traffic congestion. In a European study by Stemerding et al. the general outcome of ramp metering is that the throughput does not change (neither on the highway nor on the secondary roads) and the speed increases slightly (8% overall). The findings of the ramp metering study done by Westra et al. indicate that ramp metering can have a positive and negative impact on the travel time.

Overall the travel time increases by 2% in the morning peak. In this study, the total traveled distance was also analyzed. The distance hardly changes; neither in the morning peak nor in the evening peak. The study from Goudappel Coffeng found a 6% decrease in travel time after introducing ramp metering. At certain spots congestion can be avoided. Nevertheless, it appears that the amount of traffic in this study is so high that a good level of service (LOS) on the road cannot be achieved everywhere. A decrease in travel time of up to 48% was calculated in the study of Goudappel Coffeng. They also found an increase in the throughput by 8%. All in all the estimated effects of ramp metering are very promising.

Road trains are vehicles longer than currently allowed in Europe. The modeling exercises, both for highways, clearly show that freight can be transported using road trains without any negative effects to the speed, travel time or safety of other road users. Ludmann et al. estimates an increase in throughput of between 10% and 14% on a two-way highway. In this study, a fuel consumption reduction of 34% was calculated as well. Your country currently may have major problems with road deterioration due to heavier vehicles and overloading. The introduction of road trains is, therefore, not recommended in this country.

Bonsall et al. investigated speed limit compliance on highways. It was estimated that the throughput generally increases with between 2.9% and 5.7%.

Nevertheless, the throughput during off peak, if the speed limit is reduced by 10 km/h, will drop by 24.2%.

The first observation that needs to be made is that modeling studies have focused on the effects of ITS measures on highways. The selected sections were identified because there were congestion problems.

According to the author, there are two reasons why the focus in on highway systems. Firstly, urban ITS systems, mainly traffic management systems like SCOOT, are optimized in practice. Modeling studies with regards to these systems were carried out during the development phase, over 20 years ago. Moreover, municipalities hardly publish their experiences with urban traffic management systems. Secondly, the focus of studies on highways is due to the complexity of microscopic simulation models, which are needed to simulate ITS measures appropriately (see also Chapter 4). By reducing the research area, calibration problems etc. can be limited.

Overall it was found that the margin in speed, and more importantly travel time and throughput, is large. Speed generally drops, which indicates a safety improvement. In two cases, additional capacity is created by dedicated lanes: for dynamic road profiles and peak lanes the speed increases by almost 45% and more than 15%. The extra lane appears to create additional capacity resulting in higher throughputs and higher speeds with a lower travel time. Obviously, the travel time gains will lead to acceptance by users. In both cases, the higher speeds do not create an unsafe environment.

Shockwaves decrease by 4%, which improves the safety situation. The other exception is the Freeway Management Systems. This ITS measure shows results similar to the dedicated lane cases. Speed and throughput increase, while travel time decreases. The main aspects of the modeled freeway management system are the provision of information and improvement of safety and security. Although no shockwave information is available, better information and special attention to safety will most probably lead to a safer environment.

Travel time is generally reduced (which means a better LOS for road users). HOV lanes show a wide range of results. It needs to be kept in mind that the LOS for HOV and Single-Occupancy Vehicles (SOVs) varies. Generally, the LOS for HOV increases while the LOS for SOV decreases. In many cases, this is a policy decision that needs to be made. Unfortunately, it needs to be mentioned that the decrease in LOS for SOV will have a negative effect on user acceptance (and fuel consumption).

Throughput generally shows an increase of about 20%. A conspicuous exception is CACC with a throughput increase of 100%. A negative exception is the speed compliance system that shows a drop of up to 24% (in off peak with a reduced speed limit).

Overall it can be concluded that the modeled ITS measure clearly improves the management of traffic flows on highways and generally leads to a safer road environment.

1.4.3 INTERNATIONAL MEASURED ITS BENEFITS

The previous section summarized the findings in ex ante studies. This section focuses on the measured benefits of ITS. Data from the US, as well as Europe, appeared to be available.

The US Federal High Way Administration (FHWA) provides web-based information with regards to the benefits and costs of ITS measures. Unfortunately, the relationship between the reported benefits and costs is unclear. The benefits of a large group of projects are made available in an accessible way. These benefits are summarized in Table 1.3.

With regards to the costs, this is not the case. The cost of different realized projects, or parts thereof, are available. Accessibility of this information is less user-friendly and the link between the benefits and costs is lacking. It could be assumed that realized projects must be cost efficient. Other than implementing a pilot study to investigate effects, it would not be in a government's interest to implement non-cost-efficient projects.

With regards to Arterial Management Systems (AMS) only positive effects were measured. Red-light violations dropped (20%–70%), peak travel time (5%–11%) and fuel consumption (2%–13%) were reduced and the same LOS was provided with less rolling stock. Moreover, drivers generally felt that they were "better off".

Highway Management Systems (HMS) show very positive effects. The number of accidents is reduced by 15%–50%, injury accidents by 20%–29% and delays are 46% less. Generally, the public support the measures although throughput decreases by 14%.

The Transit Management Systems (TMS) prove to be beneficial. The vehicle on-time performance (9%–23%) and ridership (45%) increases, while the number of complaints has dropped (26%). Moreover, the same LOS is provided with the same amount of rolling stock for more passengers.

The aim of incident management systems (IMS) is to provide emergency services quicker access to reduce health problems and clear the roadway faster. There is now information available with regards to the health impacts of the IMS system. The highway closure time has dropped by 55%, while the number of secondary accidents decreased by 40%. There is even a fuel efficiency gain and consumers are thankful.

Emergency Management Systems (EMS) are another way of speeding up the response time of ambulances and other emergency vehicles to decrease the negative health related impact of accidents. The notification time decreased by an incredible 200%–500%. The efficiency of ambulances increased (10%–15%), while the public indicated that they felt more secure (70%–95%).

TABLE 1.3

Measured Benefits for Infrastructure Based ITS Measures

Application	Safety	Mobility	Productivity	Efficiency	Energy and Environment	Customer Satisfaction
Arterial Management Systems	Red-light violations −20% to −70%	Peak period travel time −5% to −11%	Same LOS witless rolling stock	N/A	Fuel consumption −2% to −13%	72% of drivers feel 'better off'
Freeway Management Systems	Reduction of accident due to ramp metering −15% to −50%	Delay −46%	Variable speed limit reduces injury accidents by −20% to 29% (−US$4 mil)	Throughput −14%	N/A	Support for ramp metering 27%–86%
Transit Management Systems	N/A	On-time bus performance +9% to +23%	Increase share riding +45% (reduction operating costs US$ 0.5 mil)	Same LOs more travelers same rolling stock	N/A	Complaints 26%
Incident Management Systems	Secondary incidents On highways −40%	Highway closure time −55%	Average duration of stall incidents −8 minutes (US$ 1.4 mil/year)	N/A	Fuel consumption −4.1 gallons/year	Received hundreds of thank you' letters
Emergency Management Systems	Notification time emergency vehicles −200% to −4,500%	N/A	Ambulance efficiency +10% to +15%	N/A	N/A	70%–95% of drivers feel more secure
Electronic Payment Systems	Uncertainty about configuration accidents +48%	Delay −85%	Revenues +12% Handling −US$ 2.7 mil	Capacity +100%	Fuel −1.2 mil/year −0.35 tons of VoC/day. 0.056 tons NO_x/day	Change of behavior 20%
Traveler Information	Fatalities −3.5%	Reliability +5% to +16%	Early/late arrivals −40% (US$ 60/user/year)	No significant change	−1.5% NO_x −25% VOC	Change of behavior 66%–86%

(Continued)

TABLE 1.3 (Continued)
Measured Benefits for Infrastructure Based ITS Measures

Application	Safety	Mobility	Productivity	Efficiency	Energy and Environment	Customer Satisfaction
Crash Prevention & Safety	Truck accidents −13% Runaway rucks −24%	Delay −6.7%	N/A	N/A	Noise −97%	Useful tool.70% of truck drivers 85% of car drivers
Road Operations & Maintenance	55% of trucker alerted	Clearance time −44%	Construction costs. US$4.1 mil/year	N/A	N/A	Signs accurate and useful
Road Weather Management	Additional decrease in-vehicle speed −26%	N/A	Labor costs −4 hours/person	N/A	N/A	30% of highway maintenance staff used system
Commercial Vehicle Operations	Out-of-service −250%	Commissioning vehicles 60% faster	Credentialing costs −60% to −75%	N/A	N/A	Hazardous material drivers in favor

Electronic Payment Systems (EPS) aim to speed up the handling of cash in toll situations, therefore, decreasing delays. In general, it was found that delays decrease (85%). Moreover, because of an increase in capacity (100%) more revenue is generated (12%). Unfortunately, an increase of accidents (48%) was measured due to drivers being unfamiliar with the new configuration of toll plazas. Finally, it was found that 20% of customers changed their behavior resulting in an energy and environmental benefit.

Traveler Information systems aim to improve traffic flows by informing customers. It was found that 66%–86% of customers change their behavior, resulting in a reduction of fatalities (3.2%), a more reliable transport system (5%–16%), less early/late arrivals (40%) and a reduction in pollution.

The Crash Prevention and Safety system that was implemented for truckers, was seen as a useful tool by 70% of truck and 85% of car drivers. There was a reduction in truck related accidents (13%) and runaway trucks (24%). Moreover, delays (6.7%) and noise levels (97%) decreased.

Road Operations and Maintenance systems show clearance time is decreased (44%) and construction costs reduced. Furthermore, customers find the signs useful.

The Road Weather Management systems, warning customers of dangerous weather related situations, have proven to drop the vehicle speed by an additional 26%.

Moreover, a drop in labor costs was found. Commercial Vehicle Operations systems show a safety (250% less out-of-service), mobility (60% faster commissioning of vehicles) and productivity (reduction of credential costs of between 60% and 75%) benefit. Hazardous material drivers appear to be very much in favor of this system.

Although not included in the overview of the FHWA, it needs to be mentioned that the Washington State Department of Transport indicates on their website that about 1%–4% of drivers will move from COVs to HOVs.

In the second version of the ITS Handbook, an overview is provided of ramp metering in the US. The benefits experienced were:

- An increase in motorway (highway) capacity of between 17% and 25%,
- An increase in speed by between 16% and 62%,
- A reduction of accidents by between 24% and 50% and a reduction of injury accidents by 71%, and A reduction of pollution by 15% (CO and HC emissions).

The other measure reported on in the ITS Handbook is Adaptive Traffic Signal Control. The findings based on several implementations in Oakland and Toronto are:

- A reduction in travel time by 7%–8%,
- A reduction in the number of stops by 22%,
- A reduction in CO and HC emissions by 4%–5%, and
- A reduction in fuel consumption of 6%. ITS-UK8 summarized the British ITS experiences recently. Five case studies were described in detail. Table 1.4 summarizes the findings.

In the previous section (Section 1.4.2), international modeling exercises generally estimated benefits for the implementation of ITS. The analysis of implemented measures confirms that ITS measures, in the US and Europe, have improved the traffic situation. Moreover, the magnitude of US and European experiences are similar.

TABLE 1.4
British ITS Experiences

	Accessibility	Safety	User Acceptance	Efficiency
Glasgow red-light camera	N/A	• 67% less • Fatalities • 14% less injuries • 8% lower speed	N/A	B/C ratio of 3.2
Norfolk interactive fiber optic signs	N/A	• Speed reduction • Potential fatality reduction 21.5%	90% of users think system is a good idea	N/A
Durham road user charging	10% more pedestrians	N/A	• 70% of users think system is a good idea • 85% of visitors consider the city to be safe	90% drop in traffic levels
M25 controlled motorway	N/A	28% less injuries	60% of users believe the system has improved	15% increase in throughput
London congestion charging	30% more cyclists	Reduced accident level	• 55% of users believe congestion has been reduced • 21% of people think parking is poorer	• 30% less congestion • 14% less journey time

The only system that was implemented in the US as well as Europe (UK), is a highway management system. Safety improves by 15%–50% (US), compared to 28% (UK). In the US, the throughput decreased by 14%, while the M25 showed a 15% increase.

Without detailed knowledge of the local situation, it is impossible to explain this difference. In both cases, there is support from the public for the measures, 79%–86% (US) versus 60% (UK).

1.5 ECONOMIC EVALUATION OF INTELLIGENT TRANSPORT SYSTEMS

1.5.1 What Do International Economic Studies Indicate?

As indicated before, information about the costs of ITS measures is very limited. Two papers were found describing the applicability of multicriteria analysis and the

TABLE 1.5
International Benefit Costs Ratios

ITS Project	B/C Ratio	Comment
Incident detection	3.8	Repaying investment in a year
Intersectional signal control	3.4	Repaying investment in a few months
Area traffic control	7.6	Extending existing technology to adjacent towns
Parking management	1.7	Even for stand-alone applications
Emergency vehicle priority	0	No cost saving but faster response time (golden hour) meant fewer people required major treatment
Weight in motion	1.8	Time saving for heavy vehicles

application and limitations of cost-benefit assessment for ITS. Unfortunately, both papers offer a theoretical description of how to carry out the assessment and do not provide any empirical results on actual projects. The only international source reporting benefit/cost ratios is the ITS Handbook.

The ratio of selected ITS applications in a number of countries have been reported to be in the order of two to eight, with the higher figures relating to urban scenarios (see Table 1.5). The implementation of ITS applications, included by PIARC, aim to obtain a capacity, safety, environmental and financial benefit. Moreover, where possible, user satisfaction is included.

Thomas compares road building (the expansion of a three-lane highway to a four-lane highway) with the implementation of a highway management system. The comparison is based on US averages. Thomas indicates that an increase of 15% in capacity can be gained by investing R0.5-million (about $55,500) per kilometer road in a highway management system, versus an investment of R5-million (about $ 555,000) per kilometer road for an additional lane to gain a 33% increase in capacity.

1.6   IMPACT OF ITS ADVANCES ON THE INDUSTRY

According to the forecast of Grand View Research for the ITS Market for the period 2019–25 [2], the value of the global market is expected to have an average growth rate of 10.5% on its 2018 value of 25.5 million US dollars. With the main requirement being the delivery of information to drivers and passengers, for safety and smart mobility reasons, companies and organizations have invested millions in V2V and V2I communication systems, as well as in smart technologies that can be on-board of vehicles. According to the same study, the use of the public transport system is also expected to grow, Europe, and North America will increase safety regulations and Asia Pacific will have major government investments on ITS.

Key players from around the world are putting forces to stand ahead in the competitive market and contribute with novel systems for ticketing and parking management, for traffic supervision and management, for vehicle autonomy and safety. All these novelties consequently affect the automotive industry, which aims in better vehicles that support the advances, and indirectly the semiconductor industry that is the pillar where automation, sensors, and actuators, smart and embedded systems are standing.

The effect in the ITS industry overall is significant, starting from hardware and moving on software and communications, which in essence leads to a major digitization step.

1.6.1 INCREASING LEVELS OF HUMAN-CENTERED AUTOMATION IN ITS WILL DRIVE THE WORLDWIDE ECONOMY

No doubt, the next steps in different industrial domains will be toward higher automation levels, where machines take increasing responsibilities and nevertheless seamlessly collaborate with humans. This holds true in future industrial production systems (e.g., collaborative robots), in healthcare (e.g., assistance robots), and especially in the ITS domain (Figure 1.2).

In the latter, the roadmap toward full automation (SAE Level 4) and autonomous driving (SAE Level 5) are quite detailed.

Recent ITS research efforts focus on the application of innovative techniques for highly automated driving. Nevertheless, it is to be mentioned that these innovations can be transferred to other industrial domains. Higher level of automation represents a three-dimensional challenge—organizational, functional, and cyber-physical (Figure 1.3). The speed of change is often limited by the organization "value chains" that create, manufacture, and service the systems and products.

1.6.2 EVOLUTION OF SAFETY AND RELIABILITY REQUIREMENTS WITH INCREASING AUTOMATION

Safety, reliability, robustness and availability requirements evolve toward automated and autonomous driving. Further, users will only accept and buy automated vehicles, when they trust the technology. Therefore ITS puts focus on features that allow making decisions more like humans, that is, by including cognitive skills, by applying human-centered artificial intelligence (AI), by noncausal reasoning on safe and

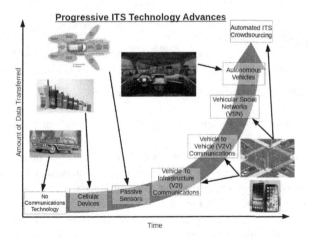

FIGURE 1.2 Cross-sectional application of ITS innovations.

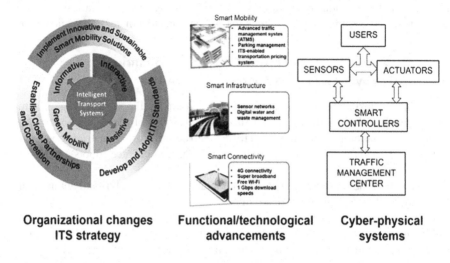

Organizational changes ITS strategy   **Functional/technological advancements**   **Cyber-physical systems**

FIGURE 1.3   A challenge in three dimensions.

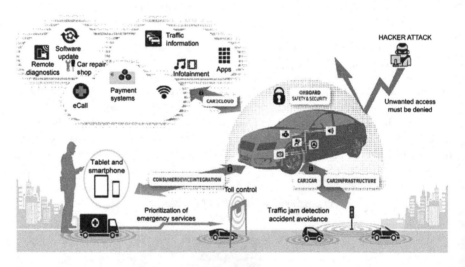

FIGURE 1.4   Reasons and use cases for connected vehicles.

reliable platforms of cost-efficient vehicles. We see this as a prerequisite to widely open the future market of automated and autonomous driving.

Despite the many required advances compared to the technical and technological state of the art, the ITS presuppose that the connected and automated vehicle will come: There are many reasons and use cases for this confidence, as indicated in Figure 1.4 in view of connected cars, and in Figure 1.5 in view of automated vehicles. No doubt, the highest benefits for society are given by a combination of automated and connected vehicles with electrification. Meanwhile, in addition to the automotive and their supplier industries, there are new players, for example Waymo, working on automated,

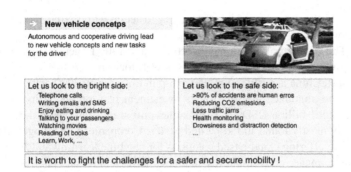

FIGURE 1.5 Advantages and use cases for automated vehicles as a subset of ITS.

even autonomous vehicles. Due to the strong importance of the automotive industry for Europe, additional endeavors are needed to keep pace with the US and Asian competition. Using cutting-edge technologies, ITS is striving to achieve real progress toward efficient, well-performing, safe, and reliable automation that is available also in critical situations, thus to earn the benefits of automation listed in Figure 1.5.

1.6.3 ITS KEY TARGETS TO STRENGTHEN EUROPEAN AND WORLDWIDE INDUSTRY AND ITS COMPETITIVENESS

ITS are associated with key industries in Europe and worldwide—the automotive industry with highly efficient automated driving and the manufacturing industries with highly automated decision processes. On the meta-level, the building blocks for both applications can effectively be assigned to three layers: on the bottom, there is the "drive" layer including motors and engine, transmission, braking, steering, and suspension (automotive)/or the processing layer (industry); the middle layer "environment sensing" is composed of on-board sensors (e.g., ultrasonic, LiDAR, radar and camera sensors, GPS, digital maps), connectivity (DSRC, WiFi, and cellular network) and data fusion (sensor data, central computing)/or the quality inspection in industrial manufacturing by inspection, measurements, etc.; on top, we find the "decision making" layer, which hosts software for decision making and the HMI to the driver/or in the industrial case the quality approval of all publications behind the United States (41%) and before China (10%) to only 13% and third place only 5 years later. Over these 5 years, China more than doubled its contributions (from 10% to 23%) to this main research venue in AI. The entire European research output published in the prime conference on AI has by now reached an alarmingly low level.

The investments made in China in AI over the last 5 years have put them into a position rivaling the United States, leaving Europe far behind. Besides the academic dimension, there are many indicators that there is an arms race for dominance in AI, with China investing heavily and continuously. Other countries including Russia, Japan, India, and South Korea have realized the importance of the field and have prioritized research moving AI forward.

In Europe, individual states and regions are investing in AI and are setting up programs to support research and development efforts. Many realize that an initiative is required. Many companies and institutions try to set up AI research and competence centers. However, in order to be competitive on a global scale, European effort is required. The regional and national efforts are generally targeted at enabling local companies to use AI, rather than moving AI itself forward.

In contrast, China is pushing for leadership in this area by 2030, with the Chinese government to announce the first major investment plan on an AI industrial park that will cost 2.1 billion USD [1]. Companies providing solutions in the area of AI are currently mostly found in the United States. Companies like Google, IBM, and Microsoft are offering cloud-based solutions for machine learning and AI. Such services include image recognition, text understanding, conversational systems, emotion detection, and more general big data analytics. Such services are extensions to conventional cloud services and are provided such that little experience is required to use them. This makes the application of AI simple, but at the same time requires to collaborate on nearly any modern and meaningful product with US companies and to share revenue with them.

Also commonly used libraries are developed and provided by companies from the United States. One prime example is TensorFlow, the open-source machine learning and AI library, which has been developed by Google's ITS Team. The efforts put in by Google, Microsoft, and IBM (and many others) and the results that are achieved are just amazing, for example, from an AI that plays GO better than any human to systems that have machine perception on a stunning level. Without a large and coordinated effort, it seems impossible to imagine to rival these technologies or to even just create technologies that are comparable in quality. The big software companies and also Amazon have moved into the AI field to extend their business models and to force a tighter integration with their customers and their other products. European companies make use of these AI services provided by US companies, as there is a clear need for these technologies and as they are not able to develop them in-house or to get them in Europe.

It is foreseeable that for many companies operating in more traditional business models, for example the automotive industry, logistics, manufacturing, and financial industries, AI will become a key success factor. Companies like Uber or Tesla, active in providing mobility, and Amazon in the area of logistics and personal assistance, invest strongly in AI as they see an opportunity to revolutionize the way they provide services to their customers. Here, too, the key players are currently in the United States, ranging from IT companies such as Google, Microsoft, and Apple to services providers such as Uber and Amazon.

At the same time, a significant part of the talent working in AI is educated in Europe and has its origins in Europe. Many graduates in computer science, machine learning, computer vision, math, and statistics get an excellent foundation at Universities across Europe on BSc, MSc, and PhD level. They are equipped with excellent skills to work in the area of AI, do research in machine learning, and create new intelligent systems. With little exciting research and development efforts in Europe and a few of the driving companies in Europe, there is a significant ITS drain. Many of the best students move after completing their degrees to work with companies such as Google, Microsoft, Amazon, Tesla, Apple, or Facebook on challenging AI problems in environments where AI is moved forward and not just applied.

In order to make Europe more attractive as a place to conduct research in AI, ITS needs to cater for the creation of an exciting research environment that is well linked to major companies. Such concerted European activity, if successful, has the clear potential not only to keep top talent in Europe but to even attract the best international students to come to Europe to do research and found new companies in the vicinity of where they find the best research.

1.6.4 GENERATING USER ACCEPTANCE BY FUNCTIONAL-SAFE ALGORITHMS AND METHODOLOGIES FOR ITS

User acceptance is essential for the successful implementation and later exploitation of any automated system. This holds true in particular for highly automated and autonomous driving as well as for the new, environment-friendly propulsion systems. In the case of autonomous vehicles, user acceptance mainly resides on the increase of driver's trust toward the self-driving automation. Trust can first be achieved through extensive testing that can be verifiable and understandable by the user. It can also be established through the good performance of the vehicle over time, which must ensure driver's and passengers' safety and comfort in all conditions that the self-driving module is active. Finally, in order to maximize the acceptance of each individual driver, it is also helpful to provide personalized systems that adapt to individual driving behaviors. It is important to understand that when the driving responsibility passes from the driver to the vehicle controller and the decisions differ significantly from those of the driver, the result can be user disappointment and disapproval of the technology.

Since autonomous driving systems use AI algorithms for predicting forthcoming situations and make better decision making, it is important that as many situations as possible have been used at training time and that the deployed technology guarantees continuous learning, not only from a single user but from all the users of this technology [5]. The correct estimation of the behaviors of surrounding vehicles will provide the autonomous vehicle controller the information needed to understand the situation better, act preventively and provide a smoother and more comfortable reaction in case of an incident. It can also help in reducing energy consumption by avoiding unnecessary throttling.

In the case of intelligent multimodal transportation and smart mobility in the context of smart cities, the ITS must guarantee informed and justified decisions and gradually prove to the users that all the recommendations help them in gaining time or reaching their destinations in a convenient manner. These directly noticeable advantages will increase people's trust and will accelerate user acceptance.

Future vehicles will have multiple energy sources and sinks for propulsion. These need to form a multiredundant, safe, and reliable systems. Requirements increase even more with the integration of cognitive intelligence: In order to behave like a human-driven vehicle, future automated vehicles will need to "look ahead" not just for potential obstacles but also on weather, terrain, and other parameters. In this context, information from external of the vehicle is to be integrated, requiring safe and secure communication. Of course, all need to be realized in a cost- and power-efficient way. We must not need several kilowatts of power just to run the signal processing of an automated vehicle. To handle all these requirements, ITS puts emphasis on the development of AI-optimized hardware (also called silicon-born AI) and on the realization of powerful, safe, reliable, and secure hardware platforms.

In terms of the underlying technology and the manufacturers behind them, AI has attracted interest from every corner of the technology world. This has ranged from graphical processor unit (GPU) and CPU companies to FPGA firms, custom ASIC markers, and more. There is a need for inference at the edge, inference at the cloud, and training in the cloud—AI processing at every level, served by a variety of processors. The importance of embedded hardware and mainly microprocessors is obvious for the AI-powered vehicles' industry. Since the continuous training of machine learning models relies on the fast processing of heterogeneous data and requires significant computing power, the leading tech companies and AI research institutions invest lots of money in researching for high-performance processors that can handle the large computation load at the edge, thus avoiding bandwidth consumption and processing bottleneck on the cloud. Typical examples are:

- The deployment of low-energy consumption, but powerful, GPUs, which can be embedded in autonomous vehicles. NVidia, a major graphics hardware accelerator developer, is currently developing AI accelerator chips that can be embedded on autonomy-level five vehicles that can process camera and other sensor input data and employ pretrained models in order to take decisions in real time.
- The design of AI computation-specific hardware that can further accelerate data processing partially "at the edge" and partially "at the cloud" in a transparent manner to the end-user. Cloud service providers such as Google and Amazon hardware divisions are working on AI accelerator chips and architectures such as the tensor processing unit (TPU). TPU is AI chip that offers 15–30 times computations than GPU's using 30–80 times less power.

The transparent processing of data, both at the edge and on the cloud, is expected to explode in the next few years, with investments on micro-chips to reach 6.5 billion USD by 2021 and the respective investment on machine learning-based knowledge inference going from zero to 1 billion USD each for data centers and edge devices per year [3]. The rise is estimated to be higher for edge devices than for the data centers, which will undertake all the inference workload leaving training—which is, in essence, the preprocessing—for the cloud backend.

In this same direction, Tesla has recently announced, at the Tesla Autonomy Day event, its new full self-driving (FSD) chip [4]. Manufactured by Samsung in Austin, TX, the custom chip, says the company, was built with autonomy and safety in mind and is currently shipping in its new models, including S, X, and 3. Two 260-mm chips feature a pair of neural network processing modules that can handle 36 trillion operations per second (TOPS) each with a power consumption of 72 W. Two such chips will be installed on each of the company's FSD computer boards, delivering 144 TOPS for collecting and processing data from radars, cameras, and ultrasonic sensors, using the embedded deep neural network architecture. The company is claiming "FSD" capability at the hardware level, for all the vehicles that are equipped with these chips. For that, says Tesla founder and CEO Elon Musk, "All you need to do is improve the software."

According to the Vice President and General Manager of Automotive Nvidia Rob Csongor, "It's not useful to compare the performance of Tesla's two-chip FSD computer against Nvidia's single-chip [Xavier] driver assistance system" but the Nvidia DRIVE AGX Pegasus computer outperforms the 144 TOPs of Tesla's chip, running at 320 TOPS and offering AI perception, localization, and path planning. Both companies agree that self-driving cars are the future of the industry and with the embedded, AI-capable, chips, and algorithms they will be able to provide safety, convenience, and efficiency at a better quality level, at the expense of computational power.

1.6.5 IMPACT ON GROWTH AND SUSTAINABILITY BY COMPLIANCE FOR INTELLIGENT TRANSPORT DECISION SYSTEMS, STANDARDS

The primary goal of ITS in this domain is to achieve cognitive decisions according to human patterns of situation awareness, perception, and decision making, based on machine perception from different kinds of sensors and data sources and AI-based learning. The processes have to follow human-centered design principles to serve and benefit humans and society as a whole, and to enhance human capabilities by the technology advances achieved toward highly automated and autonomous systems (HumanE AI Vision, Society 5.0 Vision). Compliance checking for decision systems has therefore to follow fundamental principles: technical requirements (safety, security, privacy, reliability, sustainability); human-oriented AI capabilities with a deep understanding of complex sociotechnical systems and ethical considerations.

Decision systems based on AI of the third generation are not recommended in functional safety standards at the time of writing. Particular architectures, which restrict AI-configurations to make an AI-based system safer and more predictable (no continuous machine learning, bias-free training data, guarded/monitored AI components to block unintended behavior, static neural networks validated as "black-box" element, adapted safety concepts for AI & ML) are studied and ongoing research. "Big Data" collected by (IoT) devices over time will form an essential part of AI to guide and validate decision-based systems. Explainability and accountability of machine learning methods are prerequisites to building a validation and verification environment for compliance testing of decision systems according to the fundamental principles mentioned earlier.

In the specific AI-Standardization Group ISO/IEC JTC1 SC42, there is ongoing work on AI and decision taking on a general level which should be taken into account and gives important indications what to consider for compliance checking of decision systems against high-level goals. This covers primarily safety, security and privacy issues of applied AI and decision making. Several organizations have set up ethical principles for future decision systems controlling highly automated/autonomous systems in human environments in a collaborative or noncollaborative manner. Particularly the German Ethics Report and the EC Ethics Guidelines set up generic principles to follow and a decision system must conform to the "Key Guidance for Ensuring Ethical Purpose".

ITS seeks to carefully track, support and influence as far as possible the definition and standardization of AI regulations, having clearly in focus avoiding centralizing and dramatically expanding regulation. ITS requires standards (considering safety, cybersecurity, reliability, availability, maintainability) that can be practically applied, providing guidance and enabling type approval. The AI-related regulations so far exist as additions to hardware and software products, and thus rely on the existing legislative frameworks, which lack of a concrete and detailed plan for handling AI in ITS. For example, whereas the food and drug administration provides a concrete framework for drug regulation, the US National Highway Traffic Safety Administration as part of the department of transportation issues general guidance about the operation of autonomous vehicles without defining every detail. Then it remains on national authorities to provide implementation details per case and this is mainly done for not inhibiting growth. Strict safety guidelines and consumers' and drivers' privacy are driving the development of AI solutions in ITS. The need for privacy and transparency at the same time, as it emerges from the need of insurance companies to investigate accidents, coupled with the requirements for AI transparency and accountability form a field where the automotive industry has to move on with ambiguous goals and increased challenges. The whole AI transparency framework includes developing guidelines for safety, individual privacy protection, algorithm transparency, and explainable decision making in order to turn public opinion in favor of autonomous systems and increase the public trust to them.

1.7 THE SOCIETAL IMPACT OF ITS

ITS and their latest advances (Artificial intelligence-enabled management systems, autonomous driving, electric-connected vehicles, etc.) are enormously important not only considering the benefits they bring to the economy but also to the society as a whole. The following subsections explain how ITS (with a focus on disruptive technologies) can pave the way for changes in the fundamentals of our society.

1.7.1 SMART MOBILITY

According to the Electric Vehicle Outlook report by Bloomberg New Energy Finance [6] the number of cars produced worldwide is expected to rise, and so does the number and ratio of electric vehicles, which is expected to reach 32% of the world's vehicles by 2040. Europe is the second main car manufacturer after China and it is vital for Europe to keep its strong place in the mobility market. The societal challenges for the smart mobility sector relate to the CO_2 emission reduction, improvement of the quality of air especially in urban areas, development of a sustainable mobility plan for all (including the elderly and impaired), reduction of accidents and congestion. The adoption of next-generation ITS is expected to address all of them.

The deployment of ITS will generate an impact on smart mobility and in alignment with the ERTRAC report on automated driving roadmap [7] in different ways. The social impact of ITS-related technologies concerns the following areas:

- Safety: enhanced traffic safety.
- Increased accessibility: facilitate access to city centers; reducing the time that people spend in traffic.
- Road capacity: better use of available road infrastructure.

- User comfort: increase user's degrees of freedom while activating the automated driving system.
- Environment: reduce CO_2 footprint by multiredundant, electrified propulsion.
- Social inclusion: guarantee access to new ITS for all, including people with disabilities and the elderly.

The outcomes of ITS deployment will help to realize fail-operational automated vehicles. More precisely, ITS will pave the way to achieve this, by providing the fundamentals of validated concepts for the main missing building blocks—fail-operational sensors, fail-operational sensor fusion, and fail-operational control and E/E architecture relying on human-inspired decision modeling, utilizing AI-based learning control architectures and signal abstraction as basis for cognitive decision making and noncausal reasoning.

Moreover, social impact through ITS will be achieved through carefully considering social and legal requirements for the design of ITS solutions, in order to increase user acceptance and provide a roadmap for the adoption of ITS relevant technologies. In general, ITS will contribute to:

- Increased safety primarily through the reduction of human error and therefore the number of accidents (90%–95% of accidents are due to human errors); this has the potential to make road traffic as safe as aviation or rail travel.
- Increased-road occupancy, ranging from approximately 50%–300% increase (depending on the penetration rate of connected and automated vehicles and the use of dynamic services such as ridesharing).
- Reduced perception of travel as lost time, contributing thus to reduce the "cost" of travel time.
- Increased mobility for users that currently cannot drive due to physical, mental, or age-related restrictions.
- Increased accessibility to cities from peri-urban/rural areas through the provision of automated first/last-mile services and its seamless integration with other public transport systems.

1.7.2 Enhancements in Traffic Safety

The list of human influence factors causing accidents is long. Examples are drowsiness, inattention, distraction, speeding, tailgating, and smartphone use while driving, etc. Many of these can at least be partly overcome by cognitive intelligence and human-like automated driving. The World Health Organization reports about 1.2 million deaths and 50 million injuries in road crashes annually, whereas the number of people killed annually in the United States due to drowsy driving reaches 5,000. Fatigue is a major factor for crashes, being behind one out of six fatal crashes on highways since it affects drivers' attention, reaction time, and ability to control the vehicle on emergency.

In order to confront such risks, in 2014 the European Union has issued a revision on driving licenses, which is mandatory for all member states (EU Directive 2014/85/

EU) and complies with the directions of the EU OSA Working Group concerning the drivers with obstructive sleep apnea syndrome (OSAS). OSAS is a chronic disease of the respiratory system that directly correlates with sleepiness, and which could take place frequently, during the day, in a very unpredictable manner. The disease has been included in the accident-risk diseases in the EU Directive of 2014 since it has been connected with road-traffic accidents.

According to Commission's estimates, "25,500 people lost their lives on EU roads in 2016 [...]. A further 135,000 people were seriously injured." Commissioner for Transport Violeta Bulc is worried about road fatalities and invited all stake-holders to work on reducing the number of deadly road accidents in 2020 to the half compared to 2010. In addition, the European Commission estimated that every year economic damage of 174 Billion € is caused by car accidents with human responsibilities.

Thus, making reliable and robust cost-efficient automated vehicles available to a broad range of customers offers strong potential to reduce the repairable economic and nonrepairable human damages. Highly automated vehicles also have the potential to greatly increase road capacity, reducing the time that people spend in traffic and reduce the environmental impact. In addition, productivity can go up, since people traveling in autonomous vehicles can work while being transported.

1.7.2.1   Accessibility and Capacity of Traffic

The main societal impacts of the automation that ITS introduce are related to enhancing traffic safety and improved-traffic efficiency. Research on the safety-related benefits from automated driving is still at its beginning, but the results are already very promising. According to a VTT study for Finland, traffic safety is expected to increase with the advent of automation and the respective impact on traffic flow is expected to be positive at automation level 3 and above. According to the study of the European Transport Safety Council, autonomous vehicles of level 3 will improve the throughput of the network and the traffic efficiency allowing vehicles to move faster and safer in waves of controlled speed. Even from autonomy level 2, improved safety will reduce traffic disruptions and congestion, positively affecting the transport system. Virtual patrolling is expected to predict or early detect accidents, and vehicle breakdown events, thus alerting drivers and automated vehicles that can avoid queuing and congestion. This will significantly reduce the ratio of incident-related traffic which is estimated to be one-fourth of the total traffic on road networks.

The technologies brought about by ITS target to increase efficiency in the use of automated driving on all kinds of roads, that is, motorways, rural roads, and also in the urban environment. The efficiency increase is achieved by two factors—(1) ease of access to transportation infrastructure and services built upon and (2) building new applications based on existing application verticals by sharing transportation infrastructure. It can be seen as a further means to support the digitalization of society. With half of the world's population living in large cities, and an expected rise to 70% in the next 30 years, the development of smart cities, where ITS will facilitate urban mobility is expected to boost the economic development of urban centers, which today produce 70% of the world's gross-domestic product.

1.7.2.2 Comfort and Enabling of User's Freedom

Increasing the comfort of driving mainly refers to increasing the degrees of freedom for drivers while reducing car accidents and avoiding injuries and fatalities. At the same time, it is important to reduce time to reach the destination and provide smoother mobility that avoids traffic jams and congestions. Autonomous driving can be a solution to the direction of comfortable and safe transportation without congestions and accidents. Autonomous driving presupposes the existence of holistic ITS solutions, through already developed key insights and technologies to enable the path toward autonomous driving. Autonomous vehicles must be able to detect traffic and weather conditions, identify moving objects such as other vehicles, bikes, and pedestrians and predict their trajectories at any condition (with reduced light, darkness, rain, fog, or snow). Thus, it is necessary to develop solutions that provide reliable and multiredundant perception and propulsion systems that are based on human-like control enabled by cognitive intelligence, knowledge, and noncausal reasoning.

1.7.2.3 Sustainability, Energy Efficiency, and Environment

Another dimension of societal impact concerns the reduction of mobility's environmental footprint. Not only Europe's environmental conscious societies are eagerly looking forward to the integration of clean mobility into their urban lives. The global trend for sustainability is obvious, and automated driving is moving forward driven by significant progress with attractive market-oriented cars. There is a push on many levels (global, EU, national, and organizational) to refine and implement enabling technologies and systems with the effect of fundamental change to our road transport paradigms and embracing the possibilities promised by the transition to automated vehicles. The transition phase from early adopters to the mass market is progressing, with a general and growing awareness that the underlying technology to implement automated driving is gaining a sufficient level of maturity. As mentioned earlier, ITS will harmonize the traffic flow due to foresighted-driving based on knowledge about the other traffic participants and its intelligence to generate the energy optimum speed profile.

According to the analysis of Morgan Stanley, ITSs offer the potential for more than 20% fuel savings, corresponding to 541 billion liters per year and corresponding to about 500 billion US$. 541 billion liters of saved fuel can directly be translated into corresponding savings of CO_2 and other emissions. Moreover, ITSs offer significant potential to increase road-network capacity. This potential was estimated by Morgan Stanley to amount up to 80% compared to the current status. Traffic flow harmonization in combination with knowledge about the movement and trajectories of other relevant traffic participants allows driving at higher speeds in shorter distances. This saves space and helps to reduce congestion. Further potential to reduce traffic jam could be generated by intelligent rerouting based on the traffic situation (the current status to guide all vehicles approaching a traffic jam through a narrow bypass is not the best option as this likely leads to an even worse traffic jam in the bypass). In the United States about 11 billion liters of gas are wasted yearly in traffic jams. Assuming just half of it could be saved by cooperative automated driving, enormous benefits would result, such as cost savings, fuel savings, emission reduction, reduction of wasted time, etc.

ITS is contributing to tap the high potential of vehicle automation among others by linking the two big automotive market drivers "automated driving" and "electrification" (for both of them, ITS is a basic enabler) by utilization of fail-operational perception and propulsion systems in order to achieve higher contribution to reaching the emission targets. All of the burning mobility challenges ask for accelerated introduction of electrified, connected, and automated vehicles. Acceleration of the market ramp-up of electrified vehicles is a major topic of ITS. Hybrid vehicles as well as electric vehicles with distributed propulsion, thus vehicles with several energy sources and sinks represent excellent approaches to address the previously mentioned societal challenges. Distributed propulsion systems offer redundancy possibilities. ITS considers the multiredundant propulsion systems as a key and basis for achieving the high safety, reliability, and availability requirements for vehicle automation levels 4 and 5. ITS further enriches the optimization of the multiredundant propulsion systems in view of energy saving using information from the perception system and human-inspired decision modeling.

1.7.2.4   Social Inclusion and Mobility for All

It is straightforward to think pedestrians and cyclists as vulnerable road users that must be protected from autonomous vehicles as well as "bad" drivers, or driving conditions. However, there are many more user groups, such as impaired or elderly people and young children that use the road network or move in between vehicles, especially in the urban environment. Although the increased safety procedures and vehicle mechanisms (air-bags, passive and active safety systems, etc.) have significantly reduced severe injuries and fatalities among passengers, the number of VRUs that are injured or killed in road accidents is still decreasing slowly. The antidotes, in this case, are (1) to increase drivers' awareness for VRUs and evoke a better and safer driving behavior, and (2) to improve the road infrastructure in order to early alert drivers or autonomous vehicles about neighboring or approaching VRUs.

ITS is an excellent means to support, for example, elderly and disabled people to keep their individual mobility, and to considerably reduce their accident risk while driving a car. With the results brought about by ITS, that is, human-inspired control, automated vehicles will behave like well-trained experienced human drivers, who are driving in a cautious and defensive way, thus considering potential mistakes by other traffic participants and avoiding accidents.

1.7.3   EMPLOYMENT

It is widely known that, unfortunately, manufacturing is being gradually transferred outside Europe, at the risk of outsourcing sooner or later also the R&D and losing the IP rights. The EC has made a clear statement that this negative trend has to be reversed and that investing in production sites and capabilities is necessary to secure our wealth in the future. A proper financial framework is required and already set up by the EU to support this strategy. ITS is a perfect fit for this vision and a perfectly designed action at a European level to increase R&D and to bring manufacturing back to Europe.

1.7.3.1 Educational Impact and IP Valorization in the EU

The deployment of ITS will generate manifold knowledge on which skills are required by the future digital-industry workforce. ITS will contribute to establishing a matching of skills between concrete industrial needs and academic offerings. The new methods generated in ITS (i.e., cognitive decision making, noncausal reasoning, and human-centered AI for human-inspired decision modeling), based on new scientific development and advanced technology, will increase the opportunities for graduates and Ph.D. holders to work in the mentioned industries. For sure, the concepts developed in ITS can be transferred to other industries and challenges, leading to even higher potentials for highly skilled jobs and benefits from the IP generated through the deployment of ITS.

1.7.3.2 Impact on Employment

The deployment of ITS will safeguard existing jobs and build the basis for further growth in employment in the transport and digital industry. This will help to gain technological leadership, competitiveness, and future growth within the concerned business segments.

REFERENCES

1. Cyranoski, D. (2018). China enters the battle for AI talent. *Nature*, 553(7688), 260. Available from: https://www.nature.com/articles/d41586-018-00604-6.
2. Grand View Research, (2019). Intelligent transportation system (ITS) market size, share & trends analysis report by type (ATIS, ATMS, ATPS, APTS, EMS), by application (road safety & security, public transport), and segment forecasts, 2019-2025. Grand View Research. Available from: https://www.researchandmarkets.com/r/qatx98.
3. Morgan, T. P. (2018). Inferring the future of the FPGA, and then making it. *The Next Platform*. Available from: https://www.nextplatform.com/2018/10/02/inferring-the-future-of-the-fpga-and-then-making-it/.
4. Pell, R. (2019). Tesla AI chip launch prompts performance claims exchange. eeNews Embedded. Available from: https://www.eenewsembedded.com/news/tesla-ai-chip-launch-prompts-performance-claims-exchange.
5. Shalev-Shwartz, S., Shammah, S., & Shashua, A. (2016). Safe, multi-agent, reinforcement learning for autonomous driving. NIPS Workshop on Learning, Inference and Control of Multi-Agent Systems: Dec., 2016. Barcelona, Spain.
6. Bloomberg. (2019). Electric Vehicle Outlook 2019. Available from: https://about.bnef.com/electric-vehicle-outlook/.
7. ERTRAC Working Group. (2019). Connected automated driving roadmap. European Road Transport Research Advisory Council. Available from: https://www.ertrac.org/uploads/documentsearch/id57/ERTRAC-CAD-Roadmap-2019.pdf.

2 Modeling Traffic Flows

2.1 BACKGROUND

Urban transport planning started in the United States in the 1950s with the Detroit and Chicago Transport Studies and was used to inform decision-makers about the transportation system. Urban transport planning analyzes the transport system, gives forecasts on the future performance of the system, and suggests measures to improve this performance in order to meet the desired level.

Transportation modeling is used to support decision-making in the transport planning process: decisions on the future development and management of transportation systems, especially in urban areas. Traditionally, models were used as part of an overall transportation planning process which involves a forecast of travel patterns (15–25 years) in the future and an attempt to develop a future transportation system that will work effectively. The approach in traditional planning was to predict the demand and provide the needed supply (predict and provide). As it is generated by the desire to join in activities and not by the desire just to travel, demand for travel is actually a derived demand. The transportation system provides a physical connection (supply) between activities. Traditional models, therefore, use factors influencing activity patterns (i.e., travel time and costs) to describe current demands, as well as predict changes due to planned modifications in the supply (i.e., building a new road).

Over the past couple of years, the planning focus has broadened. The new aim is to utilize the existing infrastructure (supply) optimally. ITS systems have proven that they are able to assist. The requirements with regard to transportation models have, therefore, changed. Traffic flows need to be described in a detailed, dynamic way, taking vehicle interaction into account.

This chapter provides an overview of the type of transport models, the theoretical background, a summary with regard to available microsimulation models, and an introduction to Paramics.

2.2 LEVEL OF DETAIL OF THE MODELS

Road users make different types of choices (strategic, tactical, and operational) at various moments in time. Strategic choices, such as purchasing a vehicle or making a trip, are made (long) before the road user enters the public space. Tactical decisions, such as the departure time or route choice, are generally made as the trip starts. Some tactical decisions, such as the route choice, may be changed during the trip due to information that becomes available (i.e., congestion). Operational choices, such as accelerating, decelerating, lane changes, etc., are constantly made during the trip.

Decision-makers operate on different levels. Traditionally, middle- to long-term decision-making was required. Based on that planning horizon, the four-step model (the traditional macroscopic model) was developed (see also Section 2.3). Over the years,

DOI: 10.1201/9781032691787-2

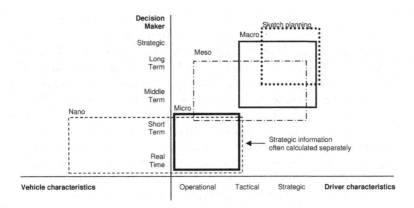

FIGURE 2.1 Trade-offs between the decision horizon and model characteristics.

the planning horizon of decision-makers changed; more strategic decisions (like changing fuel levies) were required. Tailor-made sketch planning models have been developed for that purpose. On the other hand, there has been an increasing awareness that the predict and provide philosophy is not sustainable. Ways to utilize existing capacity in a better manner is one of the new aims. ITS systems are one of the types of measures that are explored, needing real-time or short-term models. Figure 2.1 provides an overview of the type of models available.

Based on the flow and traffic dynamics representation, transport models can be divided into five types:

- Sketch planning models are, although based on the four-step transport model theory, tailor-made for specific questions. In general, a higher aggregation level is chosen. Moreover, often one or more steps are eliminated. Examples of strategic models are the Mobility Explorer and WOLOCAS (Clerx and de Vries, 1990). The Mobility Explorer uses overall characteristics (i.e., mobility per population group, car parc, and GDP) to predict long-term mobility trends. This sketch-planning model excludes the assignment step. WOLOCAS includes socio-economic and transport system data and uses the first three steps of the four-step model in a specific way to calculate additional trips from a potential settlement. WOLOCAS does not calculate a full OD-matrix, but only additional trips for one cell. As mentioned, the aim of sketch-planning models is to assist policymakers to make strategic decisions (long term) and estimate the overall (often provincial or national) effects. Sketch-planning models are static in nature.
- Macroscopic models are based on the four-step transport model. Individual vehicles are not recognized in macroscopic models. The network representation is based on links, notes, and attributes. Aspects, such as traffic controllers, are included as a note delay. Common outputs include link volumes, average speed contours, total delays, total travel times, and aggregated fuel consumption/pollution statistics. Outputs are generated for the average peak hour or work day. Changes in traffic flows in time are not included in these models. Macroscopic models are static.

- Mesoscopic models include a representation of individual vehicles (or small 'packages' of vehicles with similar characteristics). Traffic dynamics are based on fluid approximation and queuing theories. The network representation is link and lane-based (often for a corridor). Traffic control systems are detailed models based on aggregated capacity equivalents. Mesoscopic models are continuous in space (vehicles theoretically could queue on top of each other) and discrete in time (the OD-matrix is separated into small chunks; often 15-minute periods). Mesoscopic models are often referred to as semi-dynamic as they include time aspects and are, therefore, able to make a rough estimate with regards to congestion. Nevertheless, the main calculation features are static.
- Microscopic models include a representation of individual vehicles and traffic dynamics through vehicle interaction and movement. Driver behavior is included in a more detailed way (often via driver classes). Departure times of vehicles are available for every 1–5 minutes. During every calculation time step (e.g. 0.1 seconds), the position of all vehicles in the network is calculated. The outputs provide possibilities to follow vehicles, identify shock waves etc. Strategic driver information is often not included in microscopic models. Information, mostly the OD-matrix from other models (mesoscopic or macroscopic), is used as a starting point. Microscopic models are dynamic.
- Nanoscopic models are micro-simulation models that also include vehicle dynamics, such as turning radius and acceleration power. Nanoscopic models are developed for situations where microscopic models are not detailed enough. Many nanoscopic models are tailor made by car manufacturers. Nanoscopic models are also dynamic.

Due to the specific features, the models mentioned are often used for different geographical scales. Sketch-planning models have been developed to calculate national, provincial or metropole-wide changes. Macroscopic models were developed for main road networks (highway systems and other primary roads). Mesoscopic models are mostly used for corridors and include, as mentioned, traffic controller calculations, as well as secondary roads. Microscopic and nanoscopic models are generally used for any type of road or corridor where knowledge of the interaction of vehicles is needed. Generally, the research area will be smaller than for macroscopic and mesoscopic models.

2.3 MODELING PRINCIPLES

Years of experimentation and development have resulted in a general structure, which has been called the classic transport model. It resulted from practice in the 1960s and has remained more or less unaltered, despite major improvements in modeling techniques in the 1970s and 1980s (Ortúzar and Willumsen, 1994). At the outset it may be desirable to define certain terms (Davinroy et al, 1963):

- Trip generation: the determination of the number of trips associated with an area of land or other generating unit.
- Trip distribution: the determination of the interchange of a given number of trips among land areas in a region.

- Modal split: the division of trips between alternative modes of transport.
- Trip assignment: the allocation of traffic flows to the routes available between the origin and destination of a trip.

The classic transport model (macroscopic model) is presented as a sequence of four submodels (trip generation, trip distribution, modal split, and assignment). However, travel decisions are rarely taken in this sequence. A contemporary view is that the location of each sub-model in the sequence depends on the form of utility function assumed to govern travel choices.

The need to investigate congestion (and other modern life) problems, has added the requirement to model time effects and individual travel choices to the classic transport model. In the late 1980s and early 1990s, several models were developed to do so. Models adding (some) dynamics to the classic four-step model, have added disaggregate travel choice models for individuals rather than for households or zones, as well as driver behavior at vehicle level.

In the case of mesoscopic models, an attempt is made to add time in a semi-dynamic way using the theory of static model. Generally, the OD matrix is split between different departure time matrices. This provides an opportunity to obtain a rough idea of congestion problems. Microscopic models, on the other hand, have improved the assignment by making it completely dynamic. Driver choices are modeled, in small time steps, taking vehicle interactions into account. Nanoscopic models (see also Figure 2.1), operate on the same level as microscopic models, but add vehicle characteristics.

It can be concluded that the assessment of the effects of ITS systems requires the modeling of longitudinal and lateral behavior of vehicles. Sections 2.3.1 and 2.3.2 will describe the theory.

2.3.1 Longitudinal Driving Behavior

Cumming (1963) categorized the various sub-tasks that are involved in the overall driving task and paralleled the driver's role as an information processor (Rothery, 1992).

To model ITS systems requires the modeling of longitudinal (car following) and lateral behavior of vehicles.

In car following, inertia also provides direct feedback data to the driver, which is proportional to the acceleration of the vehicle. Inertia also has a smoothing effect on the performance requirements of the operator, since the large masses and limited output of drive trains eliminate high-frequency components of the task.

Car following models have not explicitly attempted to take all of these factors into account. The approach that is used assumes that a stimulus-response relationship exists that describes, at least phenomenologically, the control process of a driver-vehicle unit.

The stimulus-response equation expresses the concept that a driver of a vehicle responds to a given stimulus according to a relation (Rothery, 1992):

$$\text{Response} = \lambda \text{ Stimulus} \qquad (2.1)$$

Where:

λ is a proportionality factor.

Stimulus is composed of many factors: speed, relative speed, inter-vehicle spacing, accelerations, vehicle performance, driver thresholds (i.e., alertness, aggression) etc.

What is generally assumed in car following modeling is that a driver attempts to: (1) keep up with the vehicle ahead and (2) avoid collisions.

These two elements can be accomplished if the driver maintains a small average relative speed, Urel over short time periods, so that δt is kept small (Rothery, 1992).

$$\text{AVE} \left(U_l - U_f\right) = \text{AVE} \left(U_{rel}\right)\frac{1}{\delta t} \int_{t-\frac{\delta t}{2}}^{t+\frac{\delta t}{2}} U_{rel}(t)dt \qquad (2.2)$$

Where:

U_l is the speed of the leading vehicle.

U_f is the speed of the following vehicle.

U_{rel} is the relative speed between a lead and following vehicle.

δ is short finite time period.

This ensures that collision times are kept large and inter-vehicle spacing would not appreciably increase during the time period δt (Rothery, 1992).

$$t_c = \frac{S(t)}{U_{rel}} \qquad (2.3)$$

Where:

$S(t)$ is the inter-vehicle space.

t_c is the collision time.

The duration of δt will depend in part on alertness and the ability to estimate quantities, such as spacing, relative speed and the level of information required for the driver to assess the situation to a tolerable probability level (e.g. the probability of detecting the relative movement of an object, in this case the lead vehicle). It can also be expressed as a function of the perception time (Rothery, 1992).

Based on the driver characteristics, the relative speed should be integrated over time to reflect the recent time history of events, that is, the stimulus function should have the form of function 2.4 and be generalized so that the stimulus at a given time t, depends on the weighted sum of all earlier values of relative speed.

$$\text{AVE} \left(U_l - U_f\right) = \text{AVE} \left(U_{rel}\right)\frac{1}{\delta t} \int_{t-\frac{\delta t}{2}}^{t+\frac{\delta t}{2}} \sigma(t-t)U_{rel}(t)dt \qquad (2.4)$$

Where:

$\sigma(t)$ is a weighting function.

This function reflects a driver's estimation, evaluation and processing of earlier information (Chandler et al, 1958). The driver weights past and present information and responds at some future time (Rothery, 1992).

2.3.2   LATERAL DRIVING BEHAVIOR

Gap acceptance and lane changing are the main aspects with regard to the modeling of lateral driving behavior. Gap acceptance with regards to overtaking generally estimates if the space available in the bordering lane is enough for the vehicle to move into and back. The required space is dependent on the characteristics of the driver, the vehicle and the road. The available space depends on the characteristics of the vehicles in the bordering lane (might be oncoming) and the vehicle in front. Drivers have to perceive all these characteristics, process them and come to a decision.

Humans differ in perception capabilities, e.g. the ability to estimate distances can vary substantially between persons, and they differ in the acceptance of risk. The total acceptance process depends on many factors, of which only a subset is observable. This has led to the introduction of stochastic models (Hoogendoorn et al., undated).

The modeling of gap acceptance is done in several ways. Possible components included are the mean gap, the critical Gap, the offered gap, the accepted gap, the acceleration gap and the rejected gap. For the modeling of these gaps, a distribution is applied. An example of the distributions is provided in Table 2.1.

Mandatory lane changes, desired lane changes and overtaking movements are generally included in lane-changing models (see Figure 2.2). The desired lane change between cross-sections a and b is carried out in order to get into a better starting position for the mandatory lane change between cross-sections b and c. A driver is prepared to accept a higher risk at a mandatory lane change than at a desired lane change (Hoogendoorn et al., undated).

An overtaking, see Figure 2.3, will primarily be carried out to maintain the desired speed or to deviate less from it. The overtaking consists of two parts: a lane change to the left and a lane change to the right (assuming that driving takes place on the right-hand side). These are more or less independent maneuvers (Hoogendoorn et al, undated).

TABLE 2.1
Distribution of the Gap Acceptance Indicators

Main Gap (Class)	Distribution critical Gap	Number of offered Gaps	Number of Accepted Gaps	Distribution of Accepted Gaps	Number of Rejected Gaps	Distribution of Rejected Gaps
5	0.00	0	0	0.00	0	0.00
7	0.00	30	0	0.00	30	0.50
9	0.33	30	10	0.11	20	0.83
11	0.67	30	20	0.33	10	1.00
13	1.00	30	30	0.66	0	1.00
15	1.00	30	30	1.00	0	1.00
Sum		150	90		60	

Source: Brilon et al. (1997).

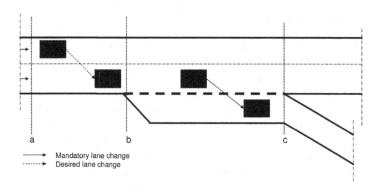

FIGURE 2.2   Mandatory and desired lane change.

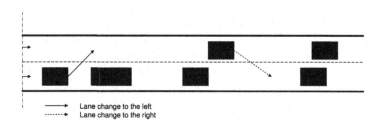

FIGURE 2.3   Lane changes at an overtaking.

As mentioned, nanoscopic models add vehicle characteristics to the dynamics included in microscopic models. These vehicle characteristics are, for example, the way the vehicle reacts in a curve. For the aim of this study, it is not necessary to add vehicle characteristics. Nanoscopic models are, therefore, not discussed any further in this thesis.

2.4   COMPARISON OF COMMERCIALLY AVAILABLE MODELS

During a European study, researchers were asked to identify which type of ITS measures should be included in modern (microscopic) models (Hugosson, et al., 1997).

The results are shown in Figure 2.4. In total, 16 measures were identified. The inclusion of traffic signals, detectors and priority for public transport, were considered most important, closely followed by ramp metering, VMS systems, incident management, dynamic route guidance and motorway flow control.

One of the widely used (over 700 locations) macroscopic models, EMME, has tried to adapt to the new requirements for modeling. Aspects, such as short- and long-term changes in transportation services, environmental impact and energy consumption, traffic restrictions or privileges (for example trucks, HOV lanes and toll roads on an urban, regional or national level) are included.

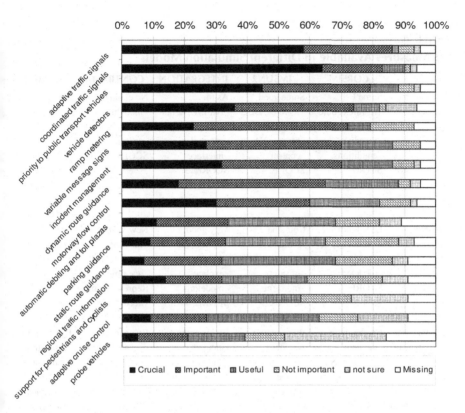

FIGURE 2.4 Overview of reported ITS-requirements for simulation tools ($n=44$).

Mesoscopic and microscopic simulation models, in which the dynamic behavior of individual agents is explicitly simulated over both time and space to generate aggregate system behavior, has been applied with increasing frequency over the past decade or more in the field of transportation systems analysis. Perhaps the best-developed application is in the area of transportation network simulation models, in which a number of operational (and often commercially supplied) software packages exist, which model second-by-second operations of individual road and/or transit vehicles over very high-fidelity representations of urban transportation networks (Miller et al, 2004). Examples (but a far from complete list) of such models include TRANSIMS (Barrett et al, 1995), Paramics (Quadstone, 2002), INTEGRATION (Aerde, van and Yager, 1988a, b), DYNASMART (Hu and Mahmassani, 1995 and Mahmassani et al., 1994), DynaMIT (Ben-Akiva et al., 1998 and Ben-Akiva et al., 1999), and VISSIM (PTV, undated).

Subsequently, much research has focused on the comparison of various ITS measures using non-traditional models (mesoscopic and microscopic simulation models). A summary table of work done by a European consortium (Schmidt and Lind, 1999) is presented in Table 2.2.

The microscopic models CORSIM, AIMSUN, and INTEGRATION, the mesoscopic model CONTRAM-I and the macroscopic model EMME/2 were compared.

TABLE 2.2

ITS Measures Available in Different Simulation Models

Measure	CORSIM	IMSUN2	INTEGRATION	CONTRAM-I	EMME/2
Adaptive urban traffic control	Some US features could be used for Swedish conditions. SCORE: 1	Some good features, DLL could be used for true VA control SCORE: 4	As for AIMSUN2 but with possible external interface SCORE: 3	No explicit modeling. Delay impact as input, SCORE: 2	No explicit modeling. Network effects if combined with microscopic model? SCORE: 2
Motorway flow control	Lane blockage but not MCS. could be modeled SCORE: 2	Could be realized, but not to the level of safety indicators. SCORE: 3	Very good, without the need for explicit modeling. SCORE: 4	New V/D function needed, Bad queuing Representation need. SCORE: 2	No queuing presentation SCORE: 1
Incident management	By changing incident duration. No rubbernecking effects. SCORE: 3	By changing incident duration. No rubbernecking effects. SCORE: 4	By changing incident duration. No rubbernecking effects. SCORE: 4	By changing capacity incl. No rubbernecking effects. SCORE: 3	By changing capacities and splitting O/D but no dynamics. SCORE: 1
Pre-trip information via radio	No explicit modeling. Departure time impact as input. SCORE: 1	No explicit modeling. Frequency of updating can be used. SCORE: 2	No explicit modeling. Capacity constraints can be used. SCORE: 2	No explicit modeling. O/D matrices for 5-minute periods can be used. SCORE: 3	No possible to modeling. SCORE: 0
Planning and evaluation of Variable Message Signs (VMS)	Does not contain a route choice model. SCORE: 0	Prepared for VMS control algorithms but route choice model dubious. SCORE: 3	Behavior and control algorithms cannot be modeling. SCORE: 2	Can be modeled by manipulation. SCORE: 2	Cannot mode reference case without information but some aspects of VMS can be modeled. SCORE: 1
Route choice effects of road pricing	No route choice.	No equilibrium assignment. No generalized cost.	Equilibrium assignment With generalized cost possible to be validated	Equilibrium assignment With generalized cost possible	Equilibrium Multi-class generalized cost assignment

(*Continued*)

TABLE 2.2 (*Continued*)
ITS Measures Available in Different Simulation Models

Measure	CORSIM	IMSUN2	INTEGRATION	CONTRAM-I	EMME/2
	SCORE: 0	SCORE: 1	SCORE: 2	SCORE: 4	SCORE: 4
Dynamic speed control	No explicit modeling.	No explicit modeling. A new ISA vehicle can be introduced.	No explicit modeling.	No possible to modeling.	No possible to modeling.
	SCORE: 1	SCORE: 3	SCORE: 1	SCORE: 0	SCORE: 0

In some instances, special features were added to the model to equip it for ITS simulations. The investigators (Schmidt and Lind, 1999) scored the different models on a scale from 0 to 5.

For the investigated models, it can be concluded that AIMSUN2, INTEGRATION, and CONTRAM-I are most equipped to model the investigated ITS measures. Classic transport models are not equipped to simulate all different ITS measures and driver behavior changes caused by ITS due to their static way of operating. Their focus is on long-term planning (although EMME/2 tries to accommodate other planning horizons as well). Based on this, it has been concluded that a macroscopic model should not be used in this book.

Other important conclusions reported by the European consortium (Schmidt and Lind, 1999) were:

- Handling of input data takes more time with microscopic simulation models than for assignment (classic) models.
- Basic behavioral models for car following etc. are good, but not sufficiently adapted to ITS.
- With proper calibration, mesoscopic and microscopic simulation models can be very useful in understanding the dynamic nature of traffic.
- It is very risky to use untried microscopic (and mesoscopic) simulation models for new areas, and when working under pressure.
- Graphical interfaces are helpful when searching for efficient input data and to understand the simulation results.
- Microscopic simulation programs are well adapted and can be recommended for traffic control measures.
- Microscopic simulation is, in principle, good for trip planning, navigation, guidance and cruise control but important modules are missing (see also requirements published by Hugosson (Hugosson, et al, 1997)).
- Macroscopic and mesoscopic assignment models are still better than dynamic models for the study of debiting systems (Schmidt and Lind, 1999).

Mesoscopic and microscopic traffic simulation models are becoming an increasingly important tool for transport systems analysis and management. They allow the traffic engineer to study and evaluate the performance of transport network systems at the tactical and operational level, under various alternative management options.

2.4.1 INPUT VARIABLES

In the previous section, a description is provided of the differences between macroscopic, mesoscopic, microscopic and nanoscopic models. It was indicated that the major differences are in the assignment step. In practice, most mesoscopic, microscopic and nanoscopic models will not carry out all four steps. Many will only calculate the last step, using input from classic four-step models or other sources, providing the required input information, including Origin-Destination (OD) information and mode choice.

Due to the dynamic aspect of microscopic models, the following information is required for the assignment step:

- Information with regards to the zones (areas of land),
- A road network that contains the physical and geometric aspects of the network (nodes, links, curves, kerb points, stop lines, ramps),
- An OD-matrix specifying the demand (segregated over time),
- Vehicle type information (private vehicles, public transport vehicles and often information with regards to the occupancy of the vehicle) providing modal split and vehicle characteristics (i.e., length),
- Driver information, including, for example, how familiar the driver is with the network, the gap acceptance, target headway and reaction time, and
- Time related information like the time steps in which the calculations need to be carried out and the percentage of traffic that will be released onto the network per time step (profiles).

2.4.2 TRANSPORT TELEMATICS INCLUDED IN MICROSCOPIC SIMULATION MODELS

The choice of microscopic simulation model will be based on the ITS measures that it is able to model. The European Consortium Smartest (1997b) carried out a thorough comparison between different microscopic simulation models. One issue investigated was the ability of models to include transport telematics (=ITS measures). The telematics functions included in their comparison are:

1. Co-ordinated traffic signals
2. Adaptive traffic signals
3. Priority or public transport vehicles
4. Ramp metering
5. Motorway flow control

6. Incident management
7. Zone access control
8. Variable message signs
9. Regional traffic information
10. Static route guidance
11. Dynamic route guidance
12. Parking guidance
13. Public transport information
14. Automatic debiting and toll plazas
15. Congestion pricing
16. Adaptive cruise control
17. Automated highway systems
18. Autonomous vehicles
19. Support for pedestrians and cyclists
20. Probe vehicles
21. Vehicle detectors

Table 2.3 provides the results of the Smartest investigation.

2.4.3 OUTPUT VARIABLES

The data that microscopic simulation models generate as a result of the calculations is generally related to different attributes. These include the following:

- Link-related information, such as volumes, average speed, and densities (per lane),
- Detector information, such as the occupancy (time that the loop is occupied by a vehicle), gap between vehicles, headway between vehicles, speed for each passing vehicle, as well as the volume and average speed, and
- Vehicle information, which includes vehicle behavior (acceleration and deceleration), the route followed travel time, etc.

More information with regard to these outputs can be found in Chapter 6. An overview of the outputs that different micro-simulation models generate is provided in Table 2.3. Table 2.4 provides the details of the developers/suppliers of these models. Table 2.5 provides an overview of the output criteria included in the models. Abbreviations have been used to summarize the information into one table.

TABLE 2.3

Transport Telematics Function Included in Microscopic Simulation Models

Model	1	2	3	4	5	6	7	8	9	10	11	12	13	14	15	16	17	18	19	20	21
AIMSUN 2	X	X		X		X	X	X		X	X			X							X
ANATOLL			X												X						X
CASIMIR		X																			X
CORSIM	X	X	X	X	X	X									X						X
DRACULA	X	X	X	X	X	X													X		X
FLEXSYT II	X	X	X	X	X	X	X							X							X
FREEVU				X	X	X														X	X
FRESIM				X	X	X															X
HUTSIM	X	X	X	X		X		X		X				X		X		X	X	X	X
INTEGRATION	X	X	X	X	X		X	X		X	X		X	X	X					X	X
MELROSE	X	X		X	X		X			X	X			X		X	X	X		X	X
MICROSIM		X								X	X										
MICSTRAN	X	X	X				X			X	X	X		X	X						X
MITSIM	X	X				X	X	X		X	X			X		X				X	X
NEMIS	X	X	X			X		X		X	X					X				X	X
PADSIM	X	X			X	X	X	X	X	X	X			X	X					X	X
PARAMICS	X	X		X		X				X							X				X
PHAROS	X		X												X						
PLANSIM-T	X	X	X											X			X			X	X

(Continued)

TABLE 2.3 (Continued)
Transport Telematics Function Included in Microscopic Simulation Models

Model	1	2	3	4	5	6	7	8	9	10	11	12	13	14	15	16	17	18	19	20	21
SHIVA																X	X	X			X
SIGSIM	X	X	X	X	X		X	X	X	X	X	X				X		X		X	X
SIMDAC			X	X	X								X			X					
SLMNET	X	X	X	X	X	X										X					X
SISTM		X	X	X	X		X				X										X
SITRA-B+	X	X	X	X		X		X		X	X	X								X	X
SITRAS	X	X	X																		X
THOREAU	X	X	X	X	X			X		X	X	X		X						X	X
VISSIM	X	X	X	X															X	X	X

TABLE 2.4
Suppliers of Microscopic Simulation Models

Model	Organization	Country
AIMSUN 2	Universitat Politèenica de Catalunya, Barcelona	Spain
ANATOLL	ISIS and Centre d'Etudes Techniques de l'Equipement	France
CASIMIR	University of New Wales, School of Civil Engineering	Australia
CORSIM	Bosch	Germany
DRACULA	Institute National de Recherche sur les Transports et la Sécurité	France
FLEXSYT II	Federal Highway Administration	USA
FREEVU	Institute for Transport Studies, University of Leeds	UK
FRESIM	Ministry of Transport	The Netherlands
HUTSIM	University of Waterloo, Department of Civil Engineering	Canada
INTEGRATION	Federal Highway Administration	USA
MELROSE	Helsinki University of Technology	Finland
MICROSIM	Queen's University, Transportation Research Group	Canada
MICSTRAN	Mitsubishi Electric Corporation	Japan
MITSIM	Centre of Parallel Computing (ZPR), University of Cologne	Germany
NEMIS	National Research Institute of Police Science	Japan
PADSIM	Massachusetts Institute of Technology	USA
PARAMICS	Mizar Automazione, Turin	Italy
PHAROS	Nottingham Trent University - NTU	UK
PLANSIM-T	The Edinburgh Parallel Computing Centre and Quadstone Ltd	UK
SHIVA	Institute for simulation and training	USA
SIGSIM	Centre of Parallel Computing (ZPR), University of Cologne	Germany
SIMDAC	ITS, University of Leeds	UK
SLMNET	Robotics Institute -CMU	USA
SISTM	University of Newcastle	UK
SITRA-B+	ONERA. Centre d'Etudes et de Recherche de Toulouse	France
SITRAS	Technical University Berlin	Germany
THOREAU	Transport Research Laboratory, Crowthorne	UK
VISSIM	ONERA - Centre d'Etudes et de Recherche de Toulouse	France

TABLE 2.5
Outputs of Different Microscopic Simulation Models

Model	Efficiency							Enrolment			Safety						Comfort		Technical Performance		Others
	E1	E2	E3	E4	E5	E6	E7	V1	V2	V3	S1	S2	S3	S4	S5	S6	F1	F2	T1	T2	
AIMSUN 2	X	X		X		X	X	X		X	X			X							
ANATOLL															X						
CASIMIR		X																			
CORSIM	X	X	X	X	X	X															
DRACULA	X	X	X												X						
FLEXSYT II	X	X	X	X	X	X	X							X					X		Number of stops
FREEVU						X															
FRESIM				X	X															X	
HUTSIM	X	X	X	X				X		X	X			X		X		X	X	X	
INTEGRATION	X	X	X	X	X	X		X		X	X		X	X	X	X				X	
MELROSE	X	X		X	X		X	X		X	X			X	X	X	X	X	X	X	Number of stops
MICROSIM	X	X								X	X										
MICSTRAN	X	X	X				X			X	X	X			X						
MITSIM	X	X	X					X		X	X			X						X	
NEMIS	X	X	X			X	X	X		X	X					X				X	
PADSIM	X	X									X			X	X	X				X	

(Continued)

TABLE 2.5 (Continued)
Outputs of Different Microscopic Simulation Models

Model	Efficiency							Enrolment					Safety				Comfort		Technical Performance		Others
	E1	E2	E3	E4	E5	E6	E7	V1	V2	V3	S1	S2	S3	S4	S5	S6	F1	F2	T1	T2	
PARAMICS	X	X		X	X	X	X	X	X	X	X						X			X	
PHAROS	X													X	X						Shock waves
PLANSIM-T	X	X	X														X			X	
SHIVA																X	X	X			Delay: Degree of saturation
SIGSIM	X	X	X	X	X		X	X	X	X	X	X				X		X		X	
SIMDAC													X			X					
SLMNET	X	X	X	X	X											X					VKT: Delay
SISTM		X	X	X	X		X				X										Time spent stopped or creeping; Amount of acceleration
SITRA-B+	X	X	X	X		X		X		X	X	X								X	
SITRAS	X	X										X									
THOREAU	X	X		X				X		X	X	X								X	
VISSIM	X	X	X	X	X									X					X	X	Transit delay due to signals: Passenger delay

The abbreviations stand for:

Efficiency	E1: Modal split
	E2: Travel time
	E3: Travel time variability
	E4: Speed
	E5: Congestion
	E6: Public transport regularity
	E7: Queue length
	V1: Exhaust emissions
	V2: Roadside pollution level
Environment	V3: Noise level
Safety	S1: Headway
	S2: Overtaking
	S3: Time-to-collision
	S4: Number of accidents
	S5: Accident speed/severity
	S6: Interaction with
	pedestrians
Comfort	F1: Physical comfort
	F2: Stress
Technical performance	T1: Fuel consumption
	T2: Vehicle operating cost

3 Sensing and Perception Systems for ITS

3.1 BACKGROUND

These systems are instrumental in enabling applications such as adaptive cruise control, lane departure warning, collision avoidance, and traffic flow optimization. Additionally, they play a crucial role in facilitating the transition toward connected and automated vehicles, where vehicles can communicate with each other and with infrastructure to improve safety and efficiency on the roads. As such, sensing and perception systems are at the forefront of innovation in the transportation industry, driving advancements that promise to revolutionize the way we move people and goods.

The ever-increasing utilization of vehicles along with the ongoing immense research in novel vehicular concepts has brought about the concept of highly automated and autonomous vehicles. The automation of vehicles—ultimately aiming at fully autonomous driving—has been identified as one major enabler to master the Grand Societal Challenges of "Individual Mobility" and "Energy Efficiency." Highly automated driving functions (ADF) are one major step to be taken. One of the major challenges to successfully realizing highly automated driving is the step from SAE Level 2 (partial automation) to SAE Level 3 (conditional automation), and above. At Level 3, the driver remains available as a fallback option in the event of a failure in the automation chain, or if the ADF reaches its operational boundaries. At higher levels (4 and 5), the driver cannot be relied upon to intervene in a timely and appropriate manner, and consequently, the automation must be capable of handling safety-critical situations on its own. This is shown in Table 3.1.

The automation of vehicles is strongly linked to their interconnection (V2V communications), as well as to their connection to the transportation (and also telecommunication) infrastructure (V2I), as those kinds of communications can pave the way for the design and delivery of innovative services and applications supporting the driver and the passengers (cooperative, connected automated mobility—CCAM). Despite the numerous advances in several initiatives related to CCAM, there are still plenty of limitations to be overcome, especially in the following areas:

1. Deployment cost reduction: At this time, CCAM solutions are associated with high costs that are associated with the distribution of the necessary infrastructure for their deployment.
2. Communication availability improvement for CCAM: Availability of state-of-the-art communication infrastructure/technologies nationwide.
3. Vehicle cooperation improvement: In-vehicle intelligence, connectivity, and coordination among heterogeneous technologies.

DOI: 10.1201/9781032691787-3

TABLE 3.1

Summary of Levels of Driving Automation for On-road Vehicles [1]

SAE Level	Name	Steering and Acceleration	Perception	Fallback	System Capabilities
		Human in Charge of Perception			
0	No automation	Driver	Driver	Driver	None
1	Driver assistance	Driver + System	Driver	Driver	Some driving modes
2	Partial automation	System	Driver	Driver	Some driving modes
		System Full in Charge of Perception			
3	Conditional automation	System	System	Driver	Some driving modes
4	High automation	System	System	System	Some driving modes
5	Full automation	System	System	System	All driving modes

4. Driving safety improvement: CCAM solutions that will assist the driver in effectively handling sudden or unforeseen situations, especially for SAE Levels 3 and beyond.
5. Business models: Solutions that will envisage new revenue generators for all involved stakeholders, that is, vehicle-to-business communications.
6. Traveler's information enhancement: Real-time, accurate, and tailored information provision to the driver, especially when information originates from multiple sources and is associated with large amounts of data.

Last, while many prototypes exist, which demonstrate CCAM technologies, they are confined to special applications and somehow limited to simple scenarios. Past and ongoing projects on CCAM focus on vehicle platooning, where vehicles operate in a well-defined and structured environment (highway scenarios).

In such a context, the vehicle needs to efficiently (in a fail-operational manner) perceive its environment, that is acquire contextual information, so as to be fully aware of its surroundings and be able to make optimal decisions regarding its velocity, direction, and overall behavior on the road.

Any mobile robot must be able to localize itself, perceive its environment, make decisions in response to those perceptions, and control actuators to move about [2]. In many ways, autonomous cars are no different. Thus many ideas from mobile robotics generally are directly applicable to highly automated (also autonomous) driving. Examples include GPS/IMU fusion with Kalman filters [3], map-based localization [4], and path planning based on trajectory scoring [5]. Actuator control for high-speed driving is different than for typical mobile robots and is very challenging. However, excellent solutions exist [6].

However, the general perception is unsolved for mobile robots and is the focus of major efforts within the research community. Perception is much more tractable within the context of autonomous driving. This is due to a number of factors. For example, the number of object classes is smaller, the classes are more distinct, rules offer a strong prior on what objects may be where at any point in time, and expensive, high-quality laser sensing is appropriate. Nevertheless, perception is still very challenging due to the extremely low acceptable error rate.

3.2 DRIVER'S SENSOR CONFIGURATIONS AND SENSOR FUSION

In driver's sensor configurations and sensor fusion, the focus is on optimizing the setup and integration of sensors to enhance the vehicle's perception capabilities and overall safety. Driven by the demand for fewer accidents and increased road safety, the automotive industry started implementing driving assistance systems into vehicles several years ago. These assistance systems include adaptive cruise control, blind-spot detection, forward collision warning, and automatic emergency braking, among others. As main sensors for monitoring the vehicle environment, 2D cameras were initially used, but in recent times, radar sensors have been increasingly employed for increased reliability. However, during the last years, it became evident that the imperfectness of capturing the vehicle environment was a major limitation, often leading to system failures or system shutdown through auto-detection. Particularly critical weather situations (snow, ice, rain, fog) and certain object properties (e.g., small-sized, non-reflecting, or transparent or mirroring obstacles) can lead to unreliable behavior. Also, mutual interference with other vehicles' active sensor units cannot be neglected with the increasing penetration of deployed assistance systems. Driving assistance represents the first level of autonomous driving. Recent research efforts address higher levels of driving autonomy (Figure 3.1 and Table 3.2), going beyond pure driver assistance systems toward fully autonomous

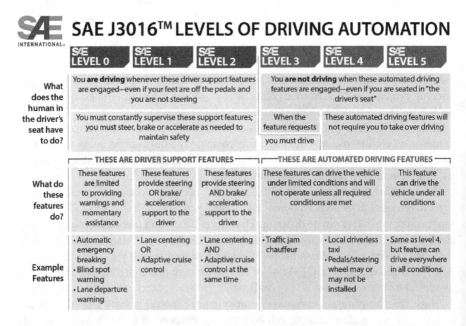

FIGURE 3.1 Evolution in ADF capabilities across SAE levels.

TABLE 3.2

Levels of Automated Driving Defined by VDA J3016 and Key Performance Figures for Autonomous Driving (Level 3+ Requires Advanced Fail-Operational Dependability and ASIL D Safety Level) [1]

Automation Level	Functional Description	Driver Interaction	Perception Redundancy	Dependability	Safety Level
Level 0	No automation	High	None	Fail-silent	QM
Level 1	Driver assistance	Medium–High	Complementary	Fail-silent	ASIL A or B
Level 2	Partial automation	Medium	Combining	Fail-safe	ASIL B
Level 3	Conditional automation	Moderate	Partially overlap	Fail-safe	ASIL C
Level 4	High automation	Seldom	Largely overlap	Fail-operational (single error)	ASIL C or D
Level 5	Full automation	None	Fully overlap	Fail-operational (single error)	ASIL D

TABLE 3.3

Forecast of OEMs Using Sensor Fusion Technologies

OEM	Expected Year for AD Level 3 Launch	Key Market	Other Sensors Fusion with LiDAR
AUDI	2017 onwards	Europe, North America, China	Stereo Camera + LRR + SRR + LiDAR
BMW	2018 onwards	Europe, North America, China	Stereo Camera + LRR + SRR
CADILLAC	2020 onwards	North America, China	Radar + Mono Camera (LiDAR expected)
FORD	2019 onwards	North America	Mono Camera + LiDAR + Radar
MERCEDEZ-BENZ	2019 onwards	Europe, North America, China	SRR + LRR + Stereo Camera + LiDAR
TESLA	2019	Europe, North America	SSR + LRR + Stereo Camera
VOLVO	2020 onwards	Europe, North America, China	SSR + LRR + Stereo Camera + LiDAR *LRR*, long-range radar; *SRR*, short range radar

driving, that is, VDA/SAE Level 3+. This involves fail-operational behavior and the highest levels of safety (ASIL D).

It is a common understanding that reliability improvement and advanced solutions for environmental perception (prerequisites for autonomous driving) can only be achieved by sensor diversity combined with data fusion approaches, due to the physical limitations of single-sensor principles.

In the automotive domain (according to all major OEMs), robust and reliable automated driving will only be achievable by combining and fusing data of three different sensor systems: LiDAR, radar, and camera, exploiting their specific strengths as depicted in Tables 3.3 and 3.4.

TABLE 3.4

Specific Strengths of Sensors

Sensor Type	Radar	LiDAR	2D Camera
Measures	Motion, velocity	3D mapping	Texture interpretation

TABLE 3.5

ADAS: Automotive Semiconductor Revenue Distribution on Device Types in 2025

Technology	Percentage	Most Important Components
Processors	37	Microprocessing units, electronic control units, digital signal processors, and systems on a chip for signal processing
Optical semiconductors	28	Complementary metal-oxide-semiconductor imaging sensors, LEDs, laser diodes, and photodetectors
Radio-frequency semiconductors	13	Radio-frequency transceivers and radar processing
Memory	12	System memory
Mixed-signal	8	Power-management integrated circuits, bus transceivers
Other	3	Discretes, other types of sensors

McKinsey predicts an overall share of 78% for processors (37%), optical (28%), and radar sensors (13%) in 2025 (Table 3.5) among automotive semiconductors, reflecting the main electronic components of highly automated vehicles as announced by OEMs.

This is evidenced not only in market reports but also in the technology roadmaps of major OEMs. Strategy Analytics has analyzed the sensor demand for environmental acquisition and indicates high annual growth rates for radar, LiDAR, and 2D camera sensors for the coming years.

However, currently available solutions for highly automated driving have not reached readiness levels suitable for the automotive industry. Although system deployment costs for these demonstration vehicles are very high, this is acceptable and normal for novel low-TRL technologies. However, the inability to achieve fail-operational levels is a significant roadblock to their adoption.

3.3 SMALL, AFFORDABLE, AND ROBUST LIDAR SENSORS WILL ENABLE HIGHLY AUTOMATED VEHICLES

LiDAR sensor technology requires the biggest push among all sensors in order to provide an economic solution. Presently, many demonstration vehicles use the HDL-64E-Laser-Scanner from Velodyne (priced at $80,000) for 360° scanning

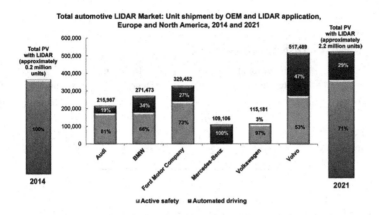

FIGURE 3.2 LiDAR market estimations for 2021.

(in good weather conditions), which is mounted on the vehicles' roof. Even smaller LiDAR modules are available, although not fully fulfill the requirements for ADAS, and are far too expensive to form a viable, scalable solution.

Therefore a low-cost $100 LiDAR sensor technology will be the major driver and enabler for robust and safe automated driving.

Figure 3.2 depicts market estimations for the LiDAR sensor technology in the year 2021. According to Frost & Sullivan, a total of approximately 2.2 million passenger vehicles implementing LiDAR sensor technology will be sold in 2021. When taking into account, for instance, BMW's automated driving strategy (which plans to implement four LiDAR sensors into every automated car), the actual number of sold LiDAR sensor systems is by factors higher than 2.2 million units per year. These estimations outline the promising market potential of automotive LiDAR technology.

In addition to a vast market potential, the positive effects of funding this research proposal are manifold for Europe's industries. This research undertaking will tackle a promising new field of research by proposing and developing innovative approaches to enable a low-cost, reliable, and automotive-qualified LiDAR sensor system. Therefore future advances will generate novel and outstanding technological know-how. As a further positive effect, the foreseen fundamental research and development activities will improve Europe's reputation as one of the world leaders in automotive and sensor technological innovation. Some key figures of state-of-the-art laser scanners are provided in Table 3.6.

Due to the required fine-grained angle and range resolutions, one of the main problems with LiDAR is that it generates a huge amount of data, which needs to be locally compressed and perfused before transmitting them to the main unit for the next level of data fusion with other sensor information. One possible approach is to provide vector object lists at the interface to reduce the high data rates, as opposed to sending raw data.

TABLE 3.6
State-of-the-Art LiDAR Specifications [7]

Sensor	Dimensional Resolution	Range	Azimuth Angle (Degree)	Accuracy	Cycle	
Quanergy M8-1	3D	150 m	360	0.05 m, -, 0.03°	33	ms
Ibeo LUX	2D	200 m	110	0.1 m, -, 0.125°	20	ms
Continental SRL1	2D	10 m	27	0.1 m, 0.5 m/s, -	10	ms
Velodyne HDL-64E S2	3D	120 m	360	0.02 m, -, 0.09°	50	ms

3.4 RADAR SENSORS AND ITS OPERATIONS

Radar sensors, a cornerstone of modern technology in transportation and vehicle safety systems, emit low-energy microwave radiation to detect objects within their vicinity. These sophisticated devices are engineered in diverse configurations to cater to specific functions. Among the key types are the Doppler radar systems, renowned for their efficiency in counting vehicles and calculating their speed through the analysis of frequency shifts in returned signals. Equally important are Frequency-Modulated Continuous Wave (FMCW) radars, which, akin to their continuous wave counterparts, provide a steady transmission power crucial for evaluating traffic flow, volume, speed, and presence. These radar sensors stand out for their precision and ease of deployment, though they are notably susceptible to environmental influences [8].

Complementing radar technologies are acoustic array sensors, which consist of an assembly of microphones tuned to identify increases in sound energy from vehicles as they approach. This acoustic detection mechanism is gradually supplanting traditional magnetic induction loops, offering a more nuanced estimation of traffic metrics such as volume, occupancy, and speed. Additionally, the utilization of road surface condition sensors, leveraging laser and infrared technologies, marks a significant stride toward enhancing traffic safety and facilitating efficient road maintenance. These sensors meticulously assess road conditions, including temperature and traction, albeit necessitating regular upkeep to ensure their effectiveness.

In the realm of parking management and vehicle identification, RFID sensors have emerged as a pivotal solution. These sensors not only streamline smart parking systems by allocating spaces but also play a critical role in the recognition and data collection of operational vehicles on the roads. Despite the broad deployment of these innovative sensors across transportation networks, challenges remain, particularly in the realms of calibration precision and system integration. According to Zhou et al. [9], Interconnection Approaches for ITS and the Automotive Society have proposed a number of communication technologies to transmit data among automobiles, transportation infrastructure, and pedestrians [10]. Their research highlights the necessity for advanced communication technologies to facilitate data exchange among vehicles, infrastructure, and pedestrians, thereby enhancing Vehicle-to-Vehicle Communication Access Technologies.

The integration of contemporary sensors, electronic frameworks, and in-vehicle communication networks into modern automobiles exemplifies the advancements in this field. These components, through a myriad of protocols and networks, enable a sophisticated interplay of data and functionality within the vehicle ecosystem. This interconnectivity not only improves the driving experience through enhanced infotainment systems but also significantly boosts vehicle safety and efficiency by providing critical data at various data rate ranges. The evolution of sensor technology and its integration into vehicular and transportation systems heralds a new era in automotive safety, efficiency, and connectivity, promising a future where vehicles not only communicate with each other but also intelligently interpret and respond to their environment.

State-of-the-art radar sensors either provide several fixed-aligned, partly overlapping beam lopes or rigid phased-array structures at the receiving front end. Typical single radar modules from Bosch are shown in Figure 3.3 (long-range radar and mid-range radar), which are quite bulky. Particularly for driving in urban environments, smart beamforming with high directionality will be necessary to properly capture the motion of other road users and outside traffic participants in the vicinity. For that reason, solutions for electronic beamforming must be developed to capture the whole vehicle environment and allow surround-vision based on radar (Figure 3.4).

Beyond advances in hardware and beamforming, the state-of-the-art programmable radar signal processing will be improved. ADAS system sensors require a latency below 500 ms. The current baseline measure for interference shielding is the different characteristics of the FMCW signal. By monitoring the radar's radio channel conditions, both, the transmitter, as well as the receiver will be upgraded to cope with increasing interferences. The state-of-the-art in the radar modules is presented

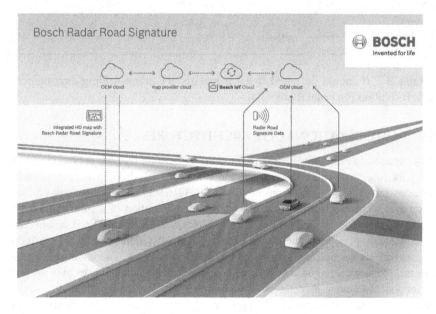

FIGURE 3.3 Bosch radar portfolio.

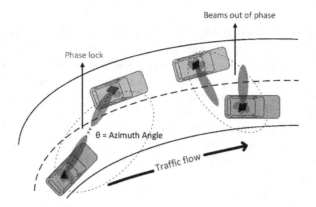

FIGURE 3.4   Situation adaptive beamforming for highly autonomous driving.

TABLE 3.7
State-of-the-Art Radar Sensor Specifications [11]

Sensor	Frequency	Bandwidth	Range	Azimuth Angle (Degree)	Accuracy	Cycle
Bosch LRR3	77 GHz	1 GHz	250 m	±15	0.1 m, 0.12 m/s, -	80 ms
Delphi ESR	77 GHz	—	174 m	±10	1.8 m, 0.12 m/s, -	50 ms
Continental ASR30x	77 GHz	1 GHz	250 m	±8.5	1.5%, 0.14 m/s, 0.1°	66 ms
SMS UMRR Type 40	24 GHz	250 GHz	250 m	±18	2.5%, 0.28 m/s, -	79 ms
TRW AC100	24 GHz	100 GHz	150 m	±8	-, -, 0.5°	—

in Table 3.7. It should be noted that particularly for urban driving environments, angle resolution is a crucial property.

3.5   FAIL-OPERATIONAL E/E ARCHITECTURES

Fail-operational E/E architectures are solutions that ensure the safe operation of highway pilot functions, valet parking, and autonomous truck driving in case individual functions fail [12]. They are required for SAE L3+ levels and must be cost-effective. These architectures guarantee the full or degraded operation of a function even if a failure occurs. Therefore, Fail-operational Electrical/Electronic (E/E) architectures are critical components in the design of safety-critical systems, particularly in industries such as automotive, aerospace, and industrial automation. These architectures are designed to ensure that systems continue to operate safely and reliably even in the event of component failures. Key aspects and considerations include redundancy at various levels, achieved through duplicate components, diverse redundant components, or functional redundancy [13]. Diverse redundancy involves using components with different designs, technologies, or suppliers to minimize common-mode failures. Additionally, fail-operational architectures incorporate fault detection and isolation mechanisms, failover mechanisms, and real-time monitoring and diagnostics to detect and respond to faults proactively [14]. Compliance with functional safety standards such as ISO 26262 in automotive systems or DO-178C/DO-254 in aerospace systems is essential, ensuring stringent safety requirements are met. Human–Machine Interfaces (HMIs) provide operators with relevant information about system status and recommended actions. Integrating and testing fail-operational architectures require rigorous validation and verification processes, including simulation and hardware-in-the-loop testing, to ensure system performance and safety under normal and fault conditions. Overall, fail-operational E/E architectures enable safety-critical systems to continue operating safely and reliably, even in the face of component failures, by incorporating redundancy, fault detection, failover mechanisms, and rigorous testing processes.

Regarding implementation, several research attempts have been taken toward the implementation of fail-operational services for ADF. The need for fail-operational behavior means that in addition to conceiving a data-flow-driven architecture capable of providing the requisite processing power for number crunching, the developed systems must also guarantee that in the event of an error (due to a sensor or hardware defect), the situation will be recognized and mitigated without impacting the vehicle's safety. A common, non-fail-operational system architecture is shown in Figure 3.5. There are remote sensor modules, which, after raw data processing and data reduction, send their data via a wired interface to a central ECU with high processing performance, where ADF main functions for environmental perception, trajectory planning, etc. are implemented. This emphasizes the importance of fail-operational architectures in ensuring continuous operation and safety in autonomous driving systems.

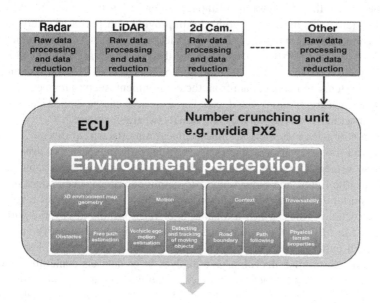

FIGURE 3.5 Non-fail-operational system architecture for autonomous driving functions.

3.6 FUSION BETWEEN SENSORS AND ARTIFICIAL INTELLIGENT

The fusion of sensors and artificial intelligence (AI) has led to significant advancements in various fields, ranging from healthcare and manufacturing to transportation and environmental monitoring. The future integration of thought inference into a car's control system represents a significant advancement made possible by the fusion of sensors and AI. By leveraging this integration, vehicles can potentially react more quickly and accurately to potential accident situations, thus enhancing overall safety on the roads. This fusion allows for the incorporation of new tools like machine learning and data fusion, which play crucial roles in enhancing ITS performance. Machine learning algorithms, for instance, can analyze vast amounts of real-time and historical transportation data to identify patterns, trends, and anomalies in driving behavior. By learning from this data, these algorithms can continuously improve their understanding of transportation dynamics and make more informed decisions. Furthermore, data fusion techniques enable the integration of data from multiple sensors and sources, providing a comprehensive view of the vehicle's surroundings and the driving environment. This integrated data can include information from cameras, LiDAR, radar, GPS, and even biometric sensors that monitor the driver's physiological state. By combining these diverse sources of data, the vehicle's control system gains a more accurate and holistic understanding of its surroundings, enabling it to anticipate and respond effectively to potential hazards or challenging driving conditions. Moreover, thought inference technology, which analyzes brain activity and patterns to infer a driver's intentions or cognitive state, can further enhance the vehicle's responsiveness and adaptability. By understanding the driver's cognitive processes and intentions, the vehicle's control system can anticipate actions or decisions before they are executed, enabling faster and more proactive responses to potential accident situations. Overall, the integration of thought inference, sensors, and AI into a car's control system holds immense promise for improving transportation safety and efficiency. By harnessing the power of machine learning and data fusion, vehicles can learn from real-world experiences and adapt their behavior accordingly, ultimately making roads safer for all users. While, Sensors can generate useful data on transportation systems and collect data from the environment, such as temperature, pressure, light intensity, motion, sound, etc. These sensors can be embedded in various devices, machines, or even worn by individuals (wearable sensors).

Fusion strategies play a crucial role in integrating data from diverse sources to analyze and forecast various conditions, such as traffic dynamics or driver behavior. These strategies involve combining data from multiple sensors, cameras, GPS devices, traffic lights, and other sources to gain a comprehensive understanding of the transportation system. One common approach to fusion strategies involves the use of association rules, which are patterns or relationships discovered within the data [10]. Machine learning techniques are then employed to identify and extract these useful patterns and trends among various traffic data sources. For example, association rule mining algorithms can analyze large datasets containing information about traffic flow, vehicle speed, road conditions, weather, and driver behavior. By identifying correlations and dependencies between different variables, these algorithms can uncover insights such as:

- Association between traffic congestion and specific times of day, weather conditions, or road construction activities.
- Relationships between driver behavior (such as aggressive driving, and lane changing patterns) and accident occurrences.
- Correlations between vehicle speed and road conditions (e.g., wet roads, icy conditions) affecting driving safety.
- Patterns in traffic flow dynamics, such as recurring congestion hotspots or peak traffic hours.

Once these patterns and associations are identified, they can be used to develop predictive models that forecast future traffic conditions, optimize traffic flow, or improve driver safety. Machine learning algorithms, including supervised learning, unsupervised learning, and reinforcement learning, are employed to train these models using historical data and real-time inputs. For instance, predictive models based on machine learning algorithms can anticipate traffic congestion in specific areas based on factors such as time of day, weather forecasts, and past traffic patterns. This information can then be used to optimize traffic signal timings, suggest alternate routes to drivers, or even adjust toll rates dynamically to manage congestion.

Developing effective learning-driven algorithms to identify and forecast traffic patterns, thereby enhancing transportation system performance, presents a significant challenge. It involves designing and implementing algorithms capable of analyzing data to perform crucial tasks: (1) data cleansing, which entails removing anomalous data collected from various sources, such as cameras and sensors, transmitted via wired or wireless links, while preserving interpretability by eliminating redundant features from the original feature space [11]; and (2) comparing and fusing data from diverse sources [10].

Ensuring the privacy and security of automobiles is paramount to reducing intrusiveness on two levels. First, integrating security and privacy protocols into communication devices supported by communication networks safeguards drivers' and passengers' privacy within the vehicular network environment. Preventing the disclosure of information that could pose a potential privacy risk is essential, considering that information sharing in a connected environment can lead to user identification. Second, optimal positioning of new devices within or outside vehicles is crucial to avoid distracting drivers while providing necessary information about various vehicle areas, thereby enhancing comfort. Careful consideration of the best interface for driver notifications, such as minimizing distractions through alarms and roadside infrastructure information, is imperative. AI can automate tasks based on human inference, contributing to distraction reduction efforts.

Integration of multiple equipment within vehicles to provide a 360° view for drivers expands detection range and precision by utilizing various information sources. However, associated hardware and software costs are substantial to achieve this objective, leading to significant data generation and processing resource requirements for alert calculations. Consequently, employing optimal processing algorithms to determine the most effective alerting interface is essential. Researchers must balance sensing and alerting device quantities while maintaining vehicle cost competitiveness to prevent potential reductions in-vehicle sales. Accelerating object detection rates and enhancing moving vehicle level and field of vision can expedite and improve responses to emergency situations [11].

In conclusion, fusion strategies that combine data from various sources and employ machine learning techniques are essential for analyzing complex transportation systems, understanding traffic dynamics, and making informed decisions to improve efficiency, safety, and overall transportation experience. In addition, the fusion of sensors and AI presents vast opportunities for innovation across various domains, ultimately leading to enhanced efficiency, accuracy, and decision-making capabilities.

3.7   ENHANCING SAFETY BASED ON ASSISTANCE APPLICATIONS SYSTEM

The assistance category application system encompasses various functionalities aimed at enhancing drivers' experiences and improving road safety. Pre-trip information applications gather data about different road conditions, offering multiple trip options for various driving routes. Parking spot locator applications assist drivers in finding available parking spaces in streets, garages, or parking lots, using technologies like magnetometers, RFID, and GPS to collect data from different parking spots. Tourist and events applications cater to travelers in unfamiliar areas by helping them locate important places in a city, find empty parking slots, and navigate routes to major events such as sporting games or concerts. These applications leverage data from sensors like radar sensors, cameras, inductive loops, and weather sensors deployed near the destination. Zhang et al. [15] discuss the importance of vehicle monitoring systems, which detect potentially dangerous situations using both built-in and outside-vehicle sensors, as well as wearable sensors on passengers. However, sensor deployment alone cannot address mobility challenges; integration with other technologies such as data analytics, automated operation tools, decision-making tools, and social and mobile networks is crucial for comprehensive information capture, analysis, and real-time sharing. One of the primary challenges faced by ITS is monitoring sensing devices' ranges on roads, vehicles, and transportation infrastructures, particularly in the presence of damaged infrastructure or environmental conditions posing threats to passengers. Careful consideration must be given to constraints, reachability, and algorithms processing sensor data to ensure timely responses to potential hazards. Combining various types of sensors, infrared and photogrammetric systems, along with efficient algorithms for multisource data fusion, can improve vehicle response times and enhance the accuracy of maps in identifying traffic and road risky conditions [10]. Data fusion techniques play a crucial role in multisensory environments, aggregating data from different sensors based on relationships such as supplementary, redundant, or cooperative data [11].

Top of Form

REFERENCES

1. ERTRAC. (2003). Automated driving roadmap. European Road Transport Research Advisory Council. https://www.ertrac.org/wp-content/uploads/2022/07/ERTRAC_Automated-Driving-2015.pdf.
2. Burgard, W., Cremers, A. B., Fox, D., Hähnel, D., Lakemeyer, G., Schulz, D., et al. (1999). Experiences with an interactive museum tour-guide robot. *Artificial Intelligence*, 114(1–2), 3–55.

3. Thrun, S., Burgard, W., & Fox, D. (2005). *Probabilistic Robotics*. MIT Press. https://doi.org/10.1016/S0004-3702(99)00070-3
4. Dellaert, F., Fox, D., Burgard, W., & Thrun, S. (1999). Monte carlo localization for mobile robots. *ICRA*, 2, 1322–1328.
5. Kelly, A., & Stentz, A. (1998). Rough terrain autonomous mobility-Part 1: a theoretical analysis of requirements. *Autonomous Robots*, 5(2), 129–161.
6. Talvala, K. L., Kritayakirana, K., & Gerdes, J. C. (2011). Pushing the limits: from lane-keeping to autonomous racing. *Annual Reviews in Control*, 35(1), 137–148.
7. de Ponte Müller, F. (2017). Survey on ranging sensors and cooperative techniques for relative positioning of vehicles. *Sensors*, 17(2), 271.
8. Orie, C. J. (2022). Sensor technologies perception for intelligent vehicle movement systems on Nigerian road network. *The Colloquium*, 10(1), 194–208
9. Zhou, Y., Dey, K. C., Chowdhury, M., & Wang, K. C. (2017). Process for evaluating the data transfer performance of wireless traffic sensors for real-time intelligent transportation systems applications. *IET Intelligent Transport Systems*, 11, 18–27.
10. Bapat, V., Kale, P., Shinde, V., Deshpande, N., & Shaligram, A. (2017). WSN application for crop protection to divert animal intrusions in the agricultural land. *Computers and Electronics in Agriculture*, 133, 88–96.
11. Ojha, T., Misra, S., & Raghuwanshi, N. S. (2017). Sensing-cloud: leveraging the benefits for agricultural applications. *Computers and Electronics in Agriculture*, 135, 96–107.
12. Oszwald, F., Obergfell, P., Traub, M., & Becker, J. (2019). Reliable fail-operational automotive E/E-architectures by dynamic redundancy and reconfiguration. *In Proceedings of the 32nd IEEE International System-on-Chip Conference (SOCC)*, Singapur, 3–6 September 2019.
13. Kohn, A., Schneider, R., Vilela, A., Roger, A., & Dannebaum, U. (2016). Architectural concepts for fail-operational automotive systems. *SAE Technical Paper.*
14. Kohn, A., Käßmeyer, M., Schneider, R., Roger, A., Stellwag, C., & Herkersdorf, A. (*2015*). Fail-operational in safety-related automotive multi-core systems. In *10th IEEE international symposium on industrial embedded systems (SIES), 8 June 2015* (pp. 1–4). IEEE, Siegen, Germany.
15. Zhang, Y., Sun, L., Song, H., &Cao, X. (2014). Ubiquitous WSN for healthcare: recent advances and future prospects. *IEEE Internet of Things Journal*, 1, 311–318.

4 Big Data Analytics for ITS

4.1 BACKGROUND

In the context of ITS, Big Data Analytics encompasses various techniques, including data mining, machine learning, predictive analytics, and visualization [1]. These techniques enable transportation agencies, operators, and planners to analyze complex datasets and uncover patterns, trends, and correlations that were previously hidden. One of the key advantages of Big Data Analytics in ITS is its ability to provide real-time insights into traffic flow, congestion patterns, and incidents [2]. By analyzing streaming data from sources such as traffic cameras, loop detectors, and GPS devices, transportation authorities can monitor traffic conditions in real time, identify congestion hotspots, and deploy timely interventions to alleviate traffic jams. Furthermore, Big Data Analytics enables predictive modeling and forecasting, allowing transportation agencies to anticipate future traffic conditions, plan infrastructure investments, and optimize resource allocation. By analyzing historical traffic data, weather patterns, and demographic trends, transportation planners can develop more accurate predictions of future traffic volumes and demand. Moreover, Big Data Analytics plays a crucial role in improving safety and security in transportation systems [3]. By analyzing data from sources such as video surveillance cameras, vehicle sensors, and incident reports, transportation agencies can identify high-risk areas, detect anomalies, and proactively address safety concerns.

One perspective on data involves viewing it as a collection of instances conforming to a particular schema, which outlines attributes and properties [4]. "Big data" refers to datasets that are exceptionally large or complex, necessitating advanced processing techniques. Sources of big data vary widely, ranging from sensors and cameras to social networks and financial transactions. In transportation systems, big data is generated by sensor devices, smart meters, remote sensing systems, and even social networks and web-based systems [5,6]. Consequently, decision support systems must handle the collection, processing, analysis, and redistribution of complex data ranging from Terabytes to Petabytes in size, a task beyond the capabilities of traditional processing systems. Efficient transportation systems rely on timely and reliable traffic data, encompassing factors like traffic load, speed, density, and composition. This data is typically collected through on-site or remote sensing techniques and then transmitted to a central repository for further processing. Different transportation modes employ various data collection techniques, involving raw data perception, optional preprocessing, and transmission to higher-level networks. Overall, Big Data Analytics holds immense potential for transforming ITS by providing actionable insights, enhancing decision-making, and ultimately improving the safety, efficiency, and sustainability of transportation systems. As the volume and complexity of transportation data continue to grow,

DOI: 10.1201/9781032691787-4

Big Data Analytics will play an increasingly critical role in shaping the future of transportation. It encompasses a wide range of topics, including data collection and management techniques for handling large volumes of transportation data from diverse sources like sensors and cameras. It involves data processing and analysis methods, including real-time processing of streaming data and advanced analytics techniques such as machine learning. Traffic prediction and modeling techniques are used to forecast traffic flow, congestion, and travel demand based on historical data and weather patterns, optimizing route planning. Incident detection and management focuses on real-time responses to traffic incidents, while transportation network optimization aims to optimize network operations and route planning. Passenger behavior analysis involves understanding preferences and satisfaction with transportation services, while safety and security analytics focus on enhancing safety measures. Infrastructure maintenance and asset management use data-driven approaches to optimize maintenance schedules, and environmental impact assessment evaluates the environmental impact of transportation systems. Finally, policy and decision support use data analytics to inform policy-making and decision-making in transportation planning

4.2 DATA COLLECTION

Data collection is a fundamental process in the domain of ITS, crucial for gathering the information needed to analyze, monitor, and optimize transportation networks. This process involves acquiring various types of data from diverse sources, including sensors, cameras, GPS devices, traffic counters, weather stations, and infrastructure systems [7]. In ITS, data collection serves multiple purposes. Firstly, it enables Traffic Monitoring by collecting data on vehicle counts, speeds, and classifications to monitor traffic flow, congestion levels, and usage patterns on roads and highways. Secondly, data collection facilitates Incident Detection by gathering information on accidents, breakdowns, and other incidents to aid rapid response and incident management by transportation authorities. Additionally, Environmental Monitoring involves gathering data on weather conditions, air quality, and other environmental factors to assess their impact on transportation operations and traveler safety. Furthermore, data collection supports Traveler Information by gathering information on travel times, congestion, and route options to provide real-time information to travelers through variable message signs, mobile apps, and websites. Lastly, Performance Evaluation is facilitated by collecting data on transportation system performance metrics, such as travel time reliability, throughput, and safety, to assess the effectiveness of transportation policies and infrastructure investments. Data collection methods in ITS vary depending on the type of information needed and the specific objectives of the transportation project. Common data collection techniques include:

- Fixed sensors: Installing stationary sensors along roadways to capture vehicle movements, speeds, and traffic flow patterns.
- Mobile sensors: Equipping vehicles with GPS devices, cameras, and other sensors to collect data on travel routes, speeds, and driving behavior.

- Remote sensing: Using remote sensing technologies, such as satellite imagery and aerial surveys, to gather data on road conditions, land use, and environmental factors.
- Surveys: Conducting surveys of travelers, businesses, and other stakeholders to gather data on travel preferences, behavior, and satisfaction with transportation services.
- Automated data collection systems: Deploying automated systems, such as toll booths, automatic license plate readers, and electronic fare collection systems, to collect data on vehicle movements and usage. Generally, effective data collection is essential for informing transportation planning, operations, and decision-making in ITS, enabling transportation agencies to improve safety, efficiency, and sustainability in the transportation network.

4.2.1   In Road Networks

Big data has emerged as a prominent research focus within the realm of ITS, evident through numerous projects worldwide. With ITS generating vast amounts of data, the advent of Big Data Analytics holds significant implications for enhancing the safety, efficiency, and profitability of transportation systems. The study of Big Data Analytics in ITS represents a thriving field, offering opportunities for innovation and improvement. "On-site" sensing techniques measure traffic data by embedding "detectors" in the road infrastructure and can be either intrusive or nonintrusive [8]. The former combines a data recorder and a sensor (e.g., pneumatic road tubes, piezoelectric sensors, magnetic loops). The sensor detects vehicles using pressure or magnetic effects and transfers this information to the data recorder on the roadside (Figure 4.1).

The main problems with such devices are their short life as they are affected by the passing of heavy vehicles and the high cost of application and maintenance. Noninvasive on-site techniques employ local range, but remote, observations. Human observers have been replaced by machines that employ technologies such as infrared, magnetic fields, microwave or acoustic radars, or video cameras. The latter is the most popular on-site nonintrusive technique, in which the road section under consideration is systematically recorded and then analyzed with special image recognition algorithms, that output the traffic load, speed, and traffic composition from video streams (Figure 4.2). On-site observation technologies have lower costs and longer life than invasive technologies but require advanced processing capabilities in the collection step and faster network connection when data processing takes place in a remote location.

Remote sensing techniques become very popular because of the rapid development of automatic positioning systems such as the Global Navigation Satellite Systems, such as GPS, Galileo, etc., that collect data from around the globe and smart mobile phones that deliver the processed information to any individual. The main principle of remote sensing techniques is that vehicles can be spotted at any time either by means of positioning devices or by Bluetooth on mobile phones and/or vehicles. The tracking of vehicles, which are considered "floating vehicles" in the road network, results in-vehicle path data (trajectory data) that depict the vehicle's positions over

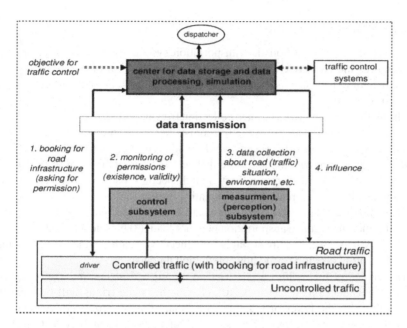

FIGURE 4.1   The structure of an ITS from the data perspective.

FIGURE 4.2   Vehicle counting using a video camera.

time. The trajectory data from many vehicles are then aggregated and used to estimate traffic density, vehicle speed, etc. [8]. When a single vehicle is equipped with
second-generation onboard diagnostics such as radars, lidars, or stereo cameras, it is
possible to act as sensor for its nearby vehicles and collect information for their speed
or number, and it is considered an extended floating Car [8]. If some of the vehicles
are connected to vehicular ad hoc networks (VANET) they can collect data from
their network and report the summary data to the aggregation layer at a higher level
[9] using interconnected base stations and access points.

4.2.2 WITHIN SMART CITIES

The growing need for a holistic transportation system within the smart city context requires the collective analysis of data from every link of the urban transport chain, including pedestrians, personal (cars, bicycles, and motors), and mass means of transportation.

The techniques employed for data collection from pedestrians and cyclist networks do not differ significantly from the techniques described previously for motorized traffic on road networks. However, they face some particular challenges that make their design and implementation more difficult and demanding. More specifically: (1) while vehicles are driven in a limited (and specific) road environment, pedestrians, and cyclists often create their own paths, (2) the limited ability to track—and distinguish between bicycles and pedestrians creates objective difficulties in automatically measuring the transportation network load, (3) bicycle and pedestrian data highly vary by season and are strongly influenced by weather conditions, thus making it difficult to develop traffic flow, prediction models.

The choice of technology to be used to record pedestrian and cyclist loads is usually dependent on two factors [10]: (1) whether pedestrians will be measured separately from the bikes or together in a "shared" or "segregated" environment and (2) whether only a temporary snapshot of pedestrian or bicycle trajectories is needed or a continuous monitoring process will be established. The recent advances in video data analysis allow the development of systems that automatically detect, count, and analyze pedestrians [9], bicycles, and other vehicle flows using existing traffic cameras [9].

The measurement of passenger traffic in mass transportation vehicles can be achieved more easily in relation to the aforementioned cases of vehicles, pedestrians, and bicycles, using the fact that passenger traffic can be counted on the admissions and discharges to and from the mass transportation vehicles at the designated stop points. Such systems are designed for improving the public transport experience [10] and either aim at providing synthetic-cumulative measurements (e.g., for a line or the whole transportation system) or quantitative-analytical measurements for a station/stop or a specific time period. An implicit measurement method can be based on ticket cancellations, at the fare collection devices located either in-vehicle or at station/stop entrances. The most advanced technique for measuring passenger traffic is the automatic passenger counter systems that count the persons embarking or disembarking a vehicle using nonintrusive detection techniques, such as infrared detection devices, thermoelectric sensing, video and image processing, etc. Finally, as in the case of vehicles, the use of mobile phones through Bluetooth devices allows the estimation of the number of passengers, but the use of these technologies is still in the experimental stage with few real applications [10].

4.2.3 IN MARITIME NETWORKS

The increasing interest of maritime companies in fleet management and control systems and solutions boosted the interest in vessel monitoring and maritime ITS.

However, it was the introduction of the automatic identification system (AIS) on vessels, in the 1990s, that brought new quality to vessel traffic services [10]. AISs are embedded in vessels and comprise a radio transmitter and receiver. The transmitter broadcasts at regular and short intervals, the vessel position and identity as well as metadata about its origin and destination port, so that the radio receivers of nearby vessels can automatically detect them and prevent collisions, and seaside stations can collect information about the routes of all the vessels at a short distance.

Although AIS has initially been designed for identifying aircraft and vessels and improving radar performance, air traffic control, and maritime and aircraft safety, it is nowadays used for many more purposes, ranging from vessel tracking [8–10] to situation awareness and vessel emission estimation. In many cases, the land or near-earth-based tracking of AIS is complemented by remote sensing data, such as those provided by radars on satellites and data from the long-range identification and tracking system for vessels. The biggest challenge with location data in maritime networks is that although vessels publicly report their positions using AIS, a network of collectors is needed to aggregate the data, remove redundancies, and forward them to a central repository, where they are combined with data from other sources and are analyzed in further. In order to maximize coverage, a network of coastal stations is responsible for the collection and preprocessing of AIS signals that sail near the seaside, and a network of earth stations aggregates image data and AIS data collected by satellites equipped with AIS receivers, as depicted in Figure 4.3.

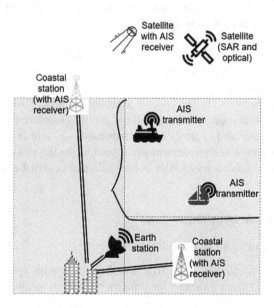

FIGURE 4.3 Maritime network surveillance using earth and satellite remote sensors for an 1D space-time diagrams.

4.2.4 In Air Transportation Networks

The global airline industry performs more than 38 million flights per year (as of 2018) and the number increased by almost 1 million per year for the last 10 years. The number of scheduled passengers surpassed four with an always increasing rate. Advances in IoT allowed collecting a massive amount of data from aircraft, from fuel consumption to aircraft maintenance and weather conditions, which can be used in order to create a better and safer customer experience. Data processing and automation are also evident in other aspects of air transportation (e.g., customer and baggage checking and handling, airport, and in-flight security and safety, etc.).

The Transportation Research Board Committee of the US Research Council (2002) envisioned small aircraft as the future of air transportation. According to the Small Aircraft Transportation System (SATS) concept, small aircraft were considered easier to pilot, more reliable, and safer but were expected to result in tens of thousands of aircraft being flown between thousands of small airports. The advent of low-cost carriers gave life to secondary airports, strengthened the vision of SATS, and transformed the air transportation networks' structure [10].

The main challenges for air transportation engineers are the alleviation of congestion and delays, the facilitation of access to air transportation services, the safety of aviation, and its environmental compatibility. There is an increasing trend in research toward the application of data mining techniques and machine learning algorithms on data collected from sensors and IoT devices for detecting flight delays.

Unmanned air vehicles (UAVs) are becoming an integral part of ITS, and are primarily engaged in remote sensing activities, such as traffic surveillance. UAVs yet are very sensitive to bad weather conditions and have limited load-carrying capacity, so they have not been widely applied to load transportation and logistics. In the transportation network surveillance scenario, data are collected from radars and sensors attached to the flying vehicles and can be from SAR images on the order of 1–2 meters to local weather data. An advantage of such systems is that they are not fixed in one position and thus can track vehicles and traffic, and thus cover wider parts of the road network. In the case of logistics, multiple UAVs can be planned to move to predefined trajectories and collaboratively complete transportation missions, always in static and controlled environments (e.g., within factory premises), but are highly sensitive to environment perturbations. In the same case but in a less constrained scenario, the use of UAVs in last-mile parcel distribution has also been envisioned [10].

4.3 BIG DATA ANALYTICS

4.3.1 Smart Cities and Road Networks

The usual way of displaying collective vehicle data is with space-time diagrams (Figure 4.3 for a 1D space example for vehicles moving in an autoroute), which can then be used to extract information such as the flow velocity at position x or at time t (determined by the slope of the respective running line), the point acceleration or deceleration at a point of time (defined by the derivative of the speed or the second

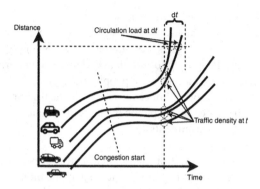

FIGURE 4.4 Maritime network surveillance using earth and satellite remote sensors or 2D space-time diagrams.

derivative of space by time), the spatial and time separation between vehicles, the traffic density, the circulation load, etc., as shown in Figure 4.4. Some of the shortcomings of remote sensing techniques are the inability of 2D space-time diagrams to show altitude information that could give extra information such as altitude accelerations, length of a line change, and the absence of a "reference point" for each vehicle, which is the actual position of the vehicle sensor system.

The classification of collection techniques can be done either based on the spatial reference level (road section, intersection, etc.) or by the method of recording the traffic data [8], followed by a detailed presentation of the traditional and modern methods of recovering these data, mainly based on the second categorization.

4.3.2 SMART CITIES AND CROWD ANALYTICS

Smart cities utilize crowd analytics as a cornerstone for optimizing urban environments and enhancing public services [9]. Crowd analytics involves the collection, processing, and analysis of data derived from various sources such as sensors, social media, and surveillance systems to understand crowd behavior, movement patterns, and demographics within urban areas. By leveraging advanced technologies such as machine learning, artificial intelligence, and Big Data Analytics, smart cities can derive valuable insights to improve urban planning, transportation systems, public safety, and resource allocation. One key application of crowd analytics in smart cities is traffic management and optimization. By analyzing real-time traffic flow data from sensors and surveillance cameras, city authorities can identify congestion hotspots, predict traffic patterns, and optimize traffic signal timings to alleviate congestion and improve overall traffic flow. Additionally, crowd analytics can inform the development of efficient public transportation systems by identifying high-demand routes, optimizing bus schedules, and enhancing passenger experience through real-time updates and personalized services.

Another important application of crowd analytics is public safety and security [8]. By analyzing data from surveillance cameras, social media, and other sources, smart cities can detect and respond to security threats, emergencies, and crowd-related incidents more effectively [10]. For example, crowd analytics can help identify suspicious behavior or unauthorized gatherings in public spaces, enabling law enforcement agencies to deploy resources proactively and prevent potential incidents. Furthermore, crowd analytics can support urban planning and development initiatives by providing insights into population density, demographic trends, and land use patterns. By understanding how people move and interact within the city, urban planners can make informed decisions about infrastructure investments, zoning regulations, and the allocation of public resources to create more livable, sustainable, and resilient urban environments.

Top of Form

Ubiquitous sensors can be part of many intelligent processing and decision support systems, ranging from tracking and monitoring applications to environment and human sensing, transportation data are one of the main data sources generated by the ubiquitous sensing of citizens in smart cities. At the same time, crowd data from other sources, such as mobile phone network data, RF and IR-based data, and satellite and video data can be exploited in order to detect traffic-related events and adjust public transportation schedules accordingly. Modern traffic management systems can also profit from crowd analytics since they can process heterogeneous data of high volume and (1) keep useful information in real-time and (2) statistical data for further analysis for longer periods.

Since the amount of crowd data that is collected from a sensor network can be huge, it is almost infeasible to process it in its entirety and to keep it collected for very long periods. In order to increase the efficiency of this process, it is necessary to filter sensor data in real-time and keep only useful summaries or values, such as hourly means and deviations, very low or very high values, etc., and feed these aggregates to a data analytics system as depicted in Figure 4.5.

As depicted in the figure, the main sources of data are citizens, who either directly—through smartphones and wearables—or indirectly—through smart vehicles, and produce geolocation data in a streamlined fashion. The data are collected and preprocessed on the edge, which means that the data collection nodes have the necessary processing power to filter out non-useful data and keep only values of interest. Such stream processing capabilities can be achieved with software explained later in this section. Only useful data are aggregated on the cloud servers, where the main processing and data archiving takes place. Using predefined rules and limits, the sensing system will be able to detect deviations and outliers in real time

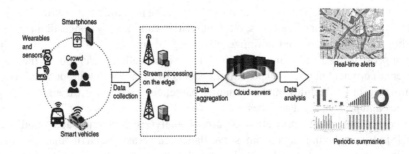

FIGURE 4.5   Crowdsensing for ITS in smart cities.

(e.g., road traffic, or passenger loads) and raise alerts. At the same time, it is possible to gather useful long-term statistics and provide the decision-makers with visualizations and summaries that can help them redesign the transportation network or the related city services. In general, crowd analytics plays a crucial role in the development of smart cities by providing valuable insights into crowd behavior, mobility patterns, and urban dynamics. By harnessing the power of data analytics, smart cities can improve public services, enhance quality of life, and create more inclusive and resilient urban communities.

4.3.3 MARITIME NETWORK ANALYTICS

The need for monitoring maritime networks and identifying the routes and positions of ships has become a necessity in the rapidly expanding maritime sector [6]. The AIS has been designed to support port authorities in achieving better maritime traffic control but soon became the key asset for companies that perform maritime transportation network analytics.

The Marine Traffic platform is the most popular vessel tracking service that collects AIS data using a worldwide network of shore-based stations. Stations collect and preprocess AIS signals (for missing values, errors, and redundancies) and forward the result to a centralized processing server for further cleaning, postprocessing, and visualization. The platform provides real-time information on the ship's position (Figure 4.6) and information on arrivals and departures from ports. A lot of data analytics projects and data mining applications have been developed on top of the historically collected data, ranging from the automatic detection of events to the prediction of vessel position and time of arrival to the port. VesselTracker [11] is a similar aggregator of AIS data that offers its registered users information about ship positions, ship metadata, reports, and statistics. The site also provides historical AIS data,

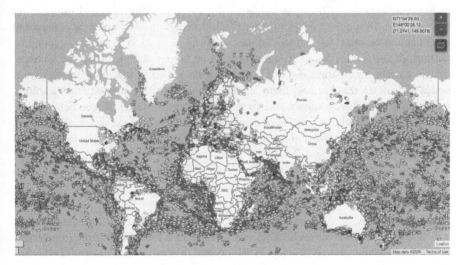

FIGURE 4.6   Real-time vessel positioning from the Marine Traffic platform.

FIGURE 4.7   An image of the traffic load for the marinas of a port.

weather forecasts, and ship alarms via email, SMS, or phone. The MariWeb platform [12] developed by IMIS Global provides tracking services for ships using a similar data collection and processing architecture.

The analysis of big maritime data can make the industry more intelligent, in a multitude of ways: first, by giving deeper insights on essential maritime operations such as route planning and fleet tracking and management, which in turn can drive more successful strategies; second, by using predictive analytics it is possible to know in advance instead of guessing and thus to improve the quality of decisions and create more successful business models; third by connecting data and information from various sources and developing a new culture of data sharing (or data business).

What can be interesting from the data analysis point of view is the detection of outliers, or simply deviations from the norm, so it is important for such applications, first to extract an abstracted naval transportation network, that contains the main nodes and routes that connect them and then to use for detecting whether a ship is moving on a previously known path or deviates from its route [7]. Another interesting output refers to the extraction of aggregate statistics about the usage of maritime infrastructures (e.g., ports and marinas) and the delays they introduce to the transportation network (Figure 4.7).

4.3.4 Aircraft Analytics

Aircraft analytics is a critical component of aviation operations, encompassing a wide range of data-driven techniques and technologies aimed at optimizing aircraft performance, safety, and efficiency. In this context, aircraft analytics involves the collection, processing, analysis, and interpretation of data from various sources, including onboard sensors, flight data recorders, maintenance logs, weather databases, air traffic control systems, and operational databases. Over the years, the air travel industry has managed to offer travelers a fast, comfortable, and affordable solution for long-range travel and thus become highly popular among travelers. This has led to an apparent increase in air traffic, which in turn led to the rise of long aircraft delays both on the ground and in the air and major economic and environmental losses. In response to growing concerns about flight optimization the use of big data collected by aircraft positioning systems and data collected from sensor systems embedded in aircraft.

In parallel, there is active research into the aviation industry to find techniques to predict flight delays with precision in order to achieve optimization of flights and minimize delays.

The process of data collection and data preprocessing does not differ from that of maritime networks. However, the main differences are in the type of analysis and the output it produces. One major type of analytics is centered around arrivals and departures in the airport, for airplanes and passengers, and is broadly described with the term "Airport Analytics" The overall aim is to reduce waiting time and this is mainly based on the ability to predict an airport's flow. Constant measuring and monitoring of passenger and plane throughput KPIs help in evaluating airport efficiency. In general, a data analytics platform that captures multiple aspects of airport operations and reports on the actual performance at any moment is the main step in developing intelligent aviation transportation systems.

Based on the same predictive analytics of passenger and airplane flow, it is possible to develop a business intelligence solution at the Carrier Company or industry level. For example, a company can have better information on the position of its aircraft, of the delays and passenger loads at the different airports, and can better schedule and plan their flights in order to maximize the number of passengers per flight, reduce expenses on fuel, minimize baggage, and passenger processing time while on the ground. When it comes to the air cargo industry, intelligent decision support systems rely on big data collected from airports, rail, and other transportation networks in order to minimize delays and empty space in cargo transportation. Overall, aircraft analytics plays a crucial role in driving operational excellence, safety, and customer satisfaction in the aviation industry, enabling airlines, aircraft manufacturers, and aviation authorities to make data-driven decisions and achieve strategic objectives effectively.

4.4 BIG DATA PROCESSING SYSTEMS FOR INTELLIGENT TRANSPORTATION

Big data processing systems are integral to the advancement of ITS, playing a crucial role in managing, analyzing, and deriving insights from the vast amounts of data generated by transportation networks. These systems leverage advanced technologies and techniques to process large volumes of diverse data sources efficiently and effectively. The main tasks that are supported by such software and are critical in big data processing are (1) data management, (2) data cleaning, (3) data aggregation, (4) data balancing, (5) data analytics, and (6) data stream processing. From their analysis, it is clear that the utilization of big data in the case of ITS can follow two different processing paths: (1) the process of streaming data and the extraction of useful information on the fly and (2) the analysis of historical data, which is aggregated on data warehouses and serves for creating business intelligence and supporting decision-making [7].

The scalability of solutions is achieved by distributing the processing load to multiple nodes and by implementing parallel processing algorithms. The two main subcomponents of the Hadoop platform [13]—the Hadoop Distributed File System (HDFS) for distributed data storage and the MapReduce framework for distributed data processing—allow parallel solutions to scale up to bigger data sizes. When a database is needed for storing data, HBase [14] is the distributed nonrelational database that operates over HDFS and adopts a simple key/value data model that allows us to scale out horizontally in distributed nodes. In the same pipeline, data querying and data analytics are served by open-source platforms such as Pig [15] and Hive [16]. More generic solutions that store data on Apache Cassandra [17] and visualize collected data using JavaScript technologies are also used to replace the data warehouse part of the architecture.

When it comes to data streams, it is important to preprocess data in real-time, instead of batch-processing data for a longer period. Apache Storm (http://storm. apache.org/) is an open-source distributed system for handling real-time data processing. The combination of Apache Kafka for data orchestration and Spark Streaming it is also possible to process live stream data. A cluster of Kafka nodes collects data from the IoT and serves it to the Spark Streaming nodes into microbatches, which can then be processed, using machine learning or AI techniques (Figure 4.8). The result of this processing is then stored back in the data warehouse for supporting data analytics. Overall, big data processing systems play a critical role in enabling intelligent transportation solutions by providing the infrastructure and capabilities to manage, analyze, and derive actionable insights from the vast amounts of data generated by transportation networks.

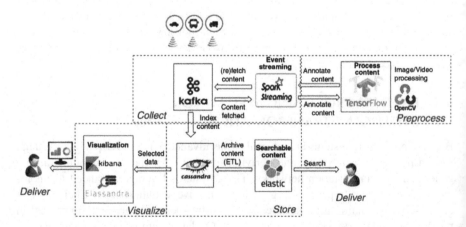

FIGURE 4.8   An overview of an architecture for stream processing and analytics of transportation data.

REFERENCES

1. Zhu, L., Yu, F. R., Wang, Y., Ning, B., & Tang, T. (2018). Big data analytics in intelligent transportation systems: a survey. *IEEE Transactions on Intelligent Transportation Systems*, 20(1), 383–398.
2. Mohamed, N., & Al-Jaroodi, J. (2014). Real-time big data analytics: applications and challenges. In *Proc. Int. Conf. High Perform. Comput. Simulation*, July 2014 (pp. 305–310).
3. Shi, Q., & Abdel-Aty, M. (2015). Big data applications in real-time traffic operation and safety monitoring and improvement on urban expressways. *Transportation Research Part C: Emerging Technologies*, 58, 380–394.
4. Alessandrini, A., Mazzarella, F., & Vespe, M. (2018). Estimated time of arrival using historical vessel tracking data. *IEEE Transactions on Intelligent Transportation Systems*, 20(1), 7–15.
5. Amini, S., Gerostathopoulos, I., & Prehofer, C. (2017). Big data analytics architecture for real-time traffic control. In *5th IEEE International Conference on Models and Technologies for Intelligent Transportation Systems (MT-ITS)* (pp. 710–715). IEEE, Naples, Italy.
6. Arguedas, V. F., Pallotta, G., & Vespe, M. (2018). Maritime traffic networks: from historical positioning data to unsupervised maritime traffic monitoring. *IEEE Transactions on Intelligent Transportation Systems*, 19(3), 722–732.
7. Grant, C., Gillis, B., & Guensler, R. (2000). Collection of vehicle activity data by video detection for use in transportation planning. *Journal of Intelligent Transportation Systems*, 5(4), 343–361.
8. Komninos, N. (2022). Transformation of industry ecosystems in cities and regions: a generic pathway for smart and green transition. *Sustainability*, 14(15), 9694.
9. Huang, H., Yao, X. A., Krisp, J. M., & Jiang, B. (2021). Analytics of location-based big data for smart cities: opportunities, challenges, and future directions. *Computers, Environment and Urban Systems*, 90, 101712.
10. Giffinger, R., & Pichler-Milanović, N. (2007). *Smart Cities: Ranking of European Medium-Sized Cities*. Vienna: Centre of Regional Science, Vienna University of Technology.
11. VesselTracker, https://www.vesseltracker.com/, 14-3-2024.
12. MariWeb Platform, https://imisglobal.com/mariweb/, 14-3-2024.
13. Hadoop Platform, https://hadoop.apache.org/, 14-3-2024.
14. HBase, https://hbase.apache.org/, 14-3-2014.
15. Pig Platform, https://pig.apache.org/, 14-3-2014.
16. Hive Platform, https://hive.apache.org/, 14-3-2014.
17. Apache Cassandra, https://cassandra.apache.org/, 14-3-2014.

5 Machine Learning in ITS

5.1 BACKGROUND

Machine learning is a collection of methods that enable computers to automate data-driven model building and programming through a systematic discovery of statistically significant patterns in the available data. While machine-learning methods are gaining popularity, the first attempt to develop a machine that mimics the behavior of a living creature was conducted by Thomas Ross in the 1930s [1]. In 1959, Arthur Samuel defined machine learning as a "Field of study that gives computers the ability to learn without being explicitly programmed" [2]. While the demonstration by Thomas Ross, then a student at the University of Washington, and his professor Stevenson Smith, included a Robot Rat that can find a way through an artificial maze [1], the study presented by Arthur Samuel included methods to program a computer "to behave in a way which, if done by human beings or animals, would be described as involving the process of learning." With the evolution of computing and communication technologies, it became possible to utilize these machine-learning algorithms to identify increasingly complex and hidden patterns in the data. Furthermore, it is now possible to develop models that can automatically adapt to bigger and more complex data sets and help decision-makers to estimate the impacts of multiple plausible scenarios in real-time.

The transportation system is evolving from a technology-driven independent system to a data-driven integrated system of systems. For example, researchers are focusing on improving existing Intelligent Transportation Systems (ITS) applications and developing new ITS applications that rely on the quality and size of the data [3]. With the increased availability of data, it is now possible to identify patterns such as the flow of traffic in real-time and the behavior of an individual driver in various traffic flow conditions to significantly improve the efficiency of existing transportation system operations and predict future trends. For example, providing real-time decision support for incident management can help emergency responders in saving lives as well as reducing incident recovery time.

Various algorithms for self-driving cars are another example of machine learning that has already begun to significantly affect the transportation system. In this case, the car (a machine) collects data through various sensors and makes driving decisions to provide a safe and efficient travel experience to passengers. In both cases, machine-learning methods search through several data sets and utilize complex algorithms to identify patterns, make decisions, and/or predict future trends.

Machine learning includes several methods and algorithms, some of which were developed before the term "machine learning" was defined and even today researchers are improving existing

DOI: 10.1201/9781032691787-5

5.2 MACHINE-LEARNING METHODS

Machine-learning methods can be characterized based on the type of "learning." There exist several basic types of learning methods, such as: (1) supervised learning where previously labeled data is used to guide the learning process; (2) unsupervised learning, where only unlabeled data is used; (3) semi-supervised learning, which uses both labeled and unlabeled data, and (4) reinforcement learning, where the learning process is guided by a series of feedback/reward cycles.

5.2.1 SUPERVISED LEARNING

The supervised learning method trains a function (or algorithm) to compute output variables based on a given data in which both input and output variables are present. For example, for a given highway, input parameters can be volume (i.e., number of vehicles per hour), current time, and age of the driver, and the corresponding output parameter can be an average traffic speed. The learning algorithm utilizes this information for automated training of a function (or algorithm) that computes the speed from a given input. Often, the goal of a learning process is to find a function that minimizes the risk of prediction error that is expressed as a difference between the real and computed output values when tested on a given data set. In such cases, the learning process can be controlled by a predetermined acceptable error threshold. The supervised learning process can be thought of as a collection of comments provided by a driving instructor during a lesson in which the instructor explains what should be done (output variables) in different situations (input variables). These comments are adapted by a student driver and turned into a driver's behavior. The predetermined thresholds can be thought of as the standards provided by external examiners such as standards published by the Department of Motor Vehicles to pass the driving exam. In this case, the student driver knows the standard way to drive (i.e., actual output) and steps to achieve it (i.e., actual inputs) before he or she starts driving lessons. For the student driver, it becomes an iterative process to achieve acceptable performance. In every iteration, the student driver makes mistakes that are corrected by the driving instructor (i.e., training the new student driver). This iterative process ends when the student successfully gets a driving license. Here, we discuss two big categories of supervised learning methods, namely, classification and regression. For example, given the speed information of individual vehicles for a highway section, the problem can be defined in the following ways:

1. Estimating how many drivers are speeding based on the speed limit provided for the highway
2. Estimating an average speed of the highway in the future based on past data. In the first case, because the solution of the problem relies on classifying the data between users who are speeding vs users who are driving below the speed limit, the problem can be thought of as a classification problem. In the second case, the solution includes mapping past data to estimate the average speed of the highway section in the future and it can be thought of as a regression function.

5.2.1.1 Classification

For a classification problem, the goal of the machine-learning algorithm is to categorize or classify given inputs based on the training data set. The training data set in a classification problem includes a set of input–output pairs categorized into classes. Many classification problems are binary, that is, only two classes such as True and False are involved. For example, the individual vehicle's speed data over time can be classified into "speeding" and "not speeding." Another example of classification is categorical classification, for example, volume and speed data over time for a highway segment can be classified into levels of service "A," "B," "C," "D," "E," and "F." When a new set of observations is presented to a trained classification algorithm, the algorithm categorizes each observation into a set of predetermined classes. Further details and selected classification methods are provided in subsequent sections.

5.2.1.2 Regression

For a regression problem, the goal of the machine-learning algorithm is to develop a relationship between outputs and inputs using a continuous function to help machines understand how outputs are changing for given inputs. The regression problems can also be envisioned as prediction problems.

For example, given the historical information about volume and speed for a given highway, the output can be the average speed of the highway for the next time period. The relationship between output variables and input variables can be defined by various mathematical functions such as linear, nonlinear, and logistic.

To summarize, supervised learning depends on the availability of historical data. It is important to note that the data must include input and corresponding known output values in order to train the model. While classification methods are used when the output is of a categorical nature, the regression methods are used for the continuous output.

5.2.2 UNSUPERVISED LEARNING

Unsupervised learning methods depend only on the underlying unlabeled data to identify hidden patterns of data instead of inferring models for known input–output pairs. Consider the same student-driver example, the learning process in this case can be thought of as the student driver with no theoretical instructions for perfect driving and he/she is driving a vehicle without the driving instructor. Without the presence of correct driving and a driving instructor, the student driver is forced to drive a vehicle by observing other drivers and deducing the correct pattern of driving. It is important to note that, the perception of "correct driving pattern" may vary for each student driver. Clustering and association are two popular families of methods for unsupervised learning problems.

5.2.2.1 Clustering

Clustering methods focus on grouping data in multiple clusters based on similarity between data points. Usually, clustering methods rely on mathematical models to identify similarities between unlabeled data points. The similarities between data points are identified by various methods such as Euclidean distance. Consider an example of a transportation engineer with a closed circuit television (CCTV) recording of peak hour traffic data for a highway segment without control information such as the speed limit of the section. The engineer is trying to identify aggressive drivers, slow drivers, and normal drivers. The engineer's goal is to find clusters such as aggressive drivers, slow drivers, and normal drivers by observing their driving patterns data such as acceleration and deceleration. In this case, it is important to note that the logic rules of such clusters are defined by the engineer based on his/her own domain expertise.

5.2.2.2 Association

The association method focuses on identifying a particular trend (or trends) in the given data set that represents major data patterns or, the so-called significant association rules that connect data patterns with each other [4]. For example, given crash data of a highway section, finding an association between the age of the drivers involved in the crash, the blood-alcohol level of the driver at the time of the crash, and time of the day can provide critical information to plan sobriety checkpoint locations and times to reduce crash as well as fatalities. For the student-driver example, the association method can be thought of as the student driver associating "normal driver behavior" with a certain age group and speed range.

In summary, unsupervised learning tries to identify complex patterns based on the logic provided in the algorithm.

5.3 UNDERSTANDING DATA

It is important to note that in both supervised and unsupervised learning, the quality, type, and size of the data are significant factors that affect the accuracy, efficiency, and robustness of the machine-learning algorithm. While the goal of any machine-learning application is to capture reality and model uncertainty, the learned model does not usually represent the real world but the reality presented by the data set.

The flowchart in Figure 5.1 presents a typical step-by-step approach for a machine-learning algorithm development process. As depicted in the figure, for any machine-learning application, the data preprocessing and learning depend on how real-world issues are defined and the data being collected.

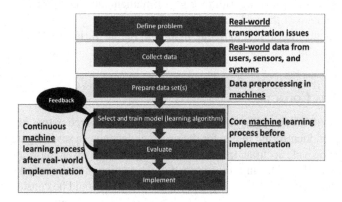

FIGURE 5.1 Machine-learning algorithm development approach.

5.3.1 PROBLEM DEFINITION

Problem definition is an important first step for any machine-learning application that directly affects all the following key steps until the model is computed. Below are the basic questions that one has to ask to define a problem and apply the appropriate machine-learning method.

- What are the input and output variables to be considered by a learning system?
- Are all of the variables of the same importance? Can we find (possibly a nonlinear weighting scheme for variables that associates each variable with some importance factor)?
- What types of machine-learning methods are we interested in (such as classification, clustering, and regression)? Do the problem and data belong to the class of (un-, semi) supervised learning methods?
- What size and type of data will be used in the learning system?

The problem identification and definition is a complex process that depends on several factors such as the general perception of the users and needs identified by decision-makers. The decision-orientated transportation planning approach is one of the methods to identify, define, and prioritize transportation planning and engineering issues [5]. For transportation data analytics applications, the problem definition provides much-needed information on the required output and possible input parameters. However, further investigation is required to justify that the solution to the problem requires a machine-learning approach. According to Tarassenko [6], to identify, whether a given problem requires a machine-learning solution or not, it must satisfy the following three criteria.

1. Given a set of input variables x and output variables y, there must be some logical evidence that a mapping between input and output variables exists such that $y5f(x)$.
2. The form of the function f in the above relationship is unknown, that is, there is no explicit algorithm or set of mathematical equations to describe the solution of the problem.
3. There must exist data to define input and output variables to train and test the learning.

5.3.2 DATA COLLECTION

The problem definition provides information about the desired output for the application; however, identifying inputs for the problem depends on several parameters. For example, to provide accurate travel time information for a given highway section, the desired output should be an average speed of the section in the near future. However, this information does not provide what input variables should be considered. Hence, the first step of data collection is to develop a list of feasible input–output variables. While there is no set of rules to develop this list, for a transportation engineer, the fundamental knowledge of transportation systems, statistics about users, and their behavior plays an important role. One of the ways to identify input variables is to develop consensus among subject matter experts via in-person or online surveys. For example, the Delphi survey method can be utilized to identify a list of input variables [7]. In this method, the experts are presented with general survey questions in the first round followed by personalized/customized questions in the second round onward to refine the opinions obtained in the first round. From the second round onward, the participants review summarized responses of the previous round before answering questions [7]. This approach ensures the compilation of a list of significant input variables for a given output. In addition, this approach can also be used to reduce the number of input variables as well as to provide weightage/ priority to each input variable for data preprocessing.

The second step of data collection is to understand how much data is sufficient. There are no specific rules to answer this question. While researchers have developed certain criteria that are algorithm-specific [8], a general approach is to select the size of representative data that captures sufficient variability of the real world for the learning algorithm to be successful most of the time.

Typically, the availability of resources, time, and educated guesses play an important role. However, as presented in Figure 5.1, it is possible to develop a feedback algorithm that evaluates and trains the learning algorithm after the real-world implementation. This approach, if executed successfully, provides a sustainable and long-term solution to overcome limited data variability and inaccuracy issues. As a general rule of thumb, it is always better to have more data than less because there exists a variety of methods that help to prioritize the information and reduce a part of it as statistically insignificant and not influential. However, these methods come with a price of increased running time and complexity of the entire system.

5.3.3 DATA FUSION

The efficiency and efficacy of machine-learning applications depend on the quality and variety of data sources considered by the learning algorithm. Data fusion methods strategically integrate and combine multiple data sets into a single, cohesive, and structured data set, extract hidden underlying knowledge, and help in improving prediction accuracy than what is possible using any of the individual data sets. In the geospatial domain, data fusion can be envisioned as combining multiple maps with various data sources. For example, combining location maps of recent highway crashes with pavement condition maps can help identify the potential impacts of deteriorating pavements on highway crashes. Providing ranking and/or priority to each data set or each input variable of each data set is one of the widely used methods to develop a structured data set. This can be achieved by decision-making algorithms [9] or by identifying weights through a survey process as explained in

Figure 5.2 presents an example of a next-generation roadway asset management system wherein several data sources can be combined using data fusion and analytics methods to predict the performance of pavements over time.

As presented in this example, the long-term pavement performance (LTPP) data which is a generalized nationwide data set can be combined with the location-specific

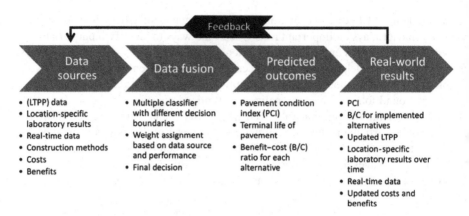

FIGURE 5.2   Intelligent pavement management system.

laboratory results and real-time data collected by various sensors. In this case, while the real-time data will receive higher weightage compared to LTPP data, the construction method and other cost-related data sets will provide a different perspective to address variability in inputs and outputs. Finally, the successful system will have the ability to predict the benefit-to-cost ratio for various alternatives to aid decisions.

Makers and the real-world results will be utilized as feedback to improve the performance of the system via the continuous machine-learning process.

Based on the data sources for input variables and their relation with output variables, the data fusion methods can be classified into the following categories [10]. A detailed review of data fusion methods is provided in Ref. [11].

- Complementary: Various input variables in various data sources can be identified as complementary when the input variables in each data source provide partial information about the solution and/or output. For example, for a given highway facility, speed information from individual cars and roadside unit locations on the highway are complementary; combining this information can improve the accuracy of speed prediction for a given facility.
- Redundant: When two input variables from two different data sources are providing the same information. For example, the radar unit and CCTV camera located on the roadside provide average speed information for the same location on the highway.
- Cooperative: When the input variables from different data sources can be combined to develop new dummy variables (i.e., information) to improve the accuracy. For example, average speed information collected from the roadway can be combined with CCTV images to estimate the density of the given facility.

5.3.4 Data Preprocessing

Data preprocessing is an essential step in which the goal is to remove noise from the raw data and convert it to a form in which the potential amount of numerical errors in complex mathematical computations will be minimized. Noise in any data set represents data patterns with errors such as measurement errors and errors due to noncalibrated data collection devices, that may significantly affect the learning process. Numerical error in this type of scientific computation is one of the most typical pitfalls [12]. Examples include truncation and rounding errors, sensitivity, conditioning, and machine precision problems. Various filtering methods can be employed to reduce or remove the noise [13].

Spatial or graphical plotting of data can often provide visual clues regarding outliers and their impact on other variables. Removing these outliers [14] may help develop better machine-learning models that fit the data. When removing outliers, a thorough understanding of the problem and fundamental knowledge of the variables involved is required to provide better quality of the final product.

An important step in data preprocessing is normalization which is a process of conditioning the data within certain boundaries to reduce redundancy, eliminate numerical problems, and improve interpretability of the results. It can be done with respect to the mean and variance of the input/output variables such that the normalized data have zero mean and unit variance as provided in the following equation:

$$\text{normalized } X_i = \frac{X_i - \mu_x}{\sigma_x} \tag{5.1}$$

where X_i is the ith data component for input (or output) variable x, and $\mu_x \, \sigma_x$ are the mean and standard deviation of variable x, respectively.

Variables can also be normalized by rescaling with respect to new minimum and maximum values, that is,

$$\text{rescaled } X_i = \frac{\left(X_i - X_{imin}\right)}{\left(X_{imax} - X_{imin}\right)} \times \left(\mu_{imax} - \mu_{imin}\right) + \mu_{imin} \tag{5.2}$$

where X_{imin} and X_{imax} are minimum and maximum values for ith component of x, and μ_{imax} respectively, and μ_{imin} are the desired maximum and minimum values of the ith component of x, respectively.

If the data values significantly differ in orders of magnitude, one may consider using logarithm transformation as a tool for normalization. Other commonly used normalization methods include square root transformation, and standardization of distributions such as Gaussian, and Poisson distributions [15].

5.4 MACHINE-LEARNING ALGORITHMS FOR DATA ANALYTICS

The type of machine-learning algorithms may vary from linear regression and classification to complex neuro-fuzzy systems. The following section presents selected popular machine-learning algorithms that can be found implemented in a variety of open-source and commercial products.

5.4.1 REGRESSION METHODS

Given a target variable (e.g., an average speed on a highway), which up to measurement errors, depends on one or several input variables (e.g., volume), regression describes

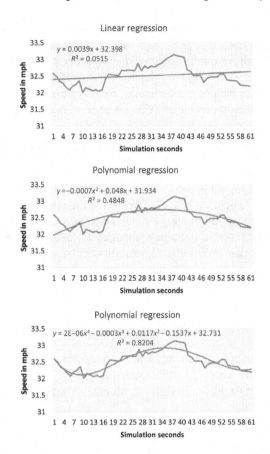

FIGURE 5.3   Examples of regression models.

the nature of dependence between the target and input variables and quantifies the error variance by finding a fitting function that maps the input variables to the target (i.e., output).

Mathematically, the training data is described as the target variables (such as speed) S_i, $i = 1,\ldots\ldots, n$ and corresponding input variables (such as volume) v_i; where each input variable can be represented as a vector of information. The general regression model is modeled by $S_i = f(v_f) + \varepsilon_i$ where ε_i is the regression error. Figure 5.3 shows examples of linear and nonlinear models.

5.4.2 DECISION TREES

A decision tree is a nonparametric method that has a structure similar to a tree or flowchart and can be utilized for classification problems [16–19] The decision tree starts with a primary question for a given problem that must be answered to solve the problem. This question is then broken down into possible solutions that may or may not have definite answers. Each possible solution is then examined by considering the result which may or may not require another decision. If the result of a possible solution requires another decision, the process continues with identifying possible

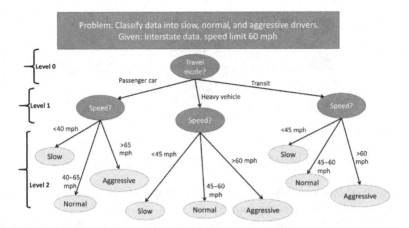

FIGURE 5.4   A simple decision tree that performs classification.

outcomes for the new decision and considering the results of each of the outcomes. The process ends when all possible decisions and outcomes are considered resulting in a tree-like structure in which the logical sequence of decisions ultimately leads to the original decision (i.e., primary question/solution of a given problem). For a classification problem, the process follows the structure of a tree where the most informative attribute to classify the data is broken down into hierarchical branches such that the next question to be asked depends on the answer to the current question.

Figure 5.4 illustrates a simple decision tree in which the problem is to classify interstate data into driver behavior (slow, normal, aggressive) based on the speed thresholds. One of the primary benefits of decision trees is that they are easy to interpret. As illustrated in the figure, from interstate traffic data if we know the speed and travel mode of any driver, we can classify that driver as a slow, normal, or aggressive driver. For example, if the driver is driving a heavy vehicle with an average speed greater than 60 mph, he or she can be classified as an aggressive driver. Breiman et al. [20] provided a general framework to construct classification and regression decision trees using the following steps given the training data set with correct labels:

1. Find the most informative attribute and the corresponding threshold to split the parent node. At first, the parent node is the root node.
2. Find the next most informative attributes, for each value of the parent node, along with their threshold to split the current node.
3. If all remaining data points at the end of the current split are of the same label, then the node becomes a leaf node with the associated label.
4. If not, either stop splitting and assign a label to the leaf node, accepting an imperfect decision, or select another attribute and continue splitting the data (and growing the tree) [20].

Furthermore, the following criteria must be considered to construct the decision tree [18].

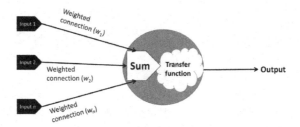

FIGURE 5.5   Neuron: a fundamental processing unit of ANN.

1. Attribute test at each level and/or node. As shown in Figure 5.5 level 1 tests the speed attribute for the data.
2. Number of splits/branching factors. In the above example, the second level is divided into three branches for each node.
3. Stopping criterion.

For a classification decision tree, the level of impurity is measured to evaluate the performance of the tree. If the decision tree classifies all data patterns into classes to which they actually belong, the splits between classes and branches are considered pure [16]. The impurity between two branches or classes can be computed based on several methods such as entropy-based impurity, Gini impurity, and misclassification impurity [18]. The following equation provides entropy-based impurity.

$$i_{\text{Entropy}}(N) = -\sum_j P(w_j)\log P(w_j) \tag{5.3}$$

where $i_{\text{Entropy}}(N)$ is the entropy of node N and $P(w_j)$ is the probability of class w_j patterns that arrive at the node N [18].

5.4.3   NEURAL NETWORKS

Neural networks (NNs) or artificial neural networks (ANNs) are designed to mimic the functions and architecture of the nervous system [21]. First, introduced in 1943 by McCulloch and Pitts [21], ANNs have gained significant popularity in the machine learning and data analytics realm. NNs are extremely powerful tools that have been applied in several transportation applications [22, 23]. Similar to the nervous system, the fundamental unit of the ANN is a neuron that utilizes the "transfer function" to calculate the output for a given input [22]. As illustrated in Figure 5.6, these neurons are connected to form a network through which data flows (i.e., input layer to other procession layers to output layer). The connections are weighted connections that scale the data flow while transfer functions in each neuron map inputs with outputs.

The general relationship between x inputs and y output can be given as:

$$y_m = f\left(B_m + \sum_{i=1}^{n} W_{im}X_i\right) \tag{5.4}$$

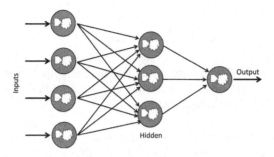

FIGURE 5.6 Typical ANN structure.

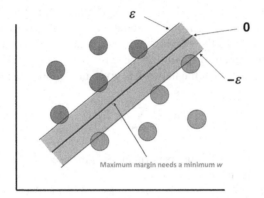

FIGURE 5.7 SVM concept.

where y is the output, X_i is ith input from the input layer X with n input variable, B_m is the bias for the neuron, W_{im} is connection weight from ith the neuron of the input layer to mth neuron. The transfer function is typically a nonlinear function such as radial basis or sigmoid.

The structure of ANN typically has three layers: input, hidden, and output. As presented in Figure 5.7, the hidden layer connects the input and output layers with an extra set of neurons. For each mth neuron in the hidden layer H, the output is given by:

$$H_m = f\left(B_{hm} + \sum_{i=1}^{n} W_{im} X_i\right) \tag{5.5}$$

where H_m is the output of mth the neuron of the hidden layer H, Bhm is the bias for the neuron in the hidden layer, x_i is ith input from the input layer X with nth input variable, and W_{im} is connection weight from ith the neuron of the input layer to mth neuron of the hidden layer.

The output from the hidden layer H becomes an input for the next layer, if there is only one hidden layer, H_m becomes an input for the output layer. Thus, prediction from the output neuron is given by:

$$P_k = f\left(B_0 + \sum_{i=1}^{n} W_{im} X_i\right) \qquad (5.6)$$

where P_k is the predicted output from kth the neuron of output layer P, B_0 is the bias for the neuron, H_i is ith input from the hidden layer H with n hidden neurons, and W_{im} is connection weight from ith neuron of the hidden layer to kth neuron of the output layer.

It is evident from these equations that weights, that connect input to hidden and hidden to output layers, are the backbone of the ANN architecture upon which the performance of the algorithm is dependent. The learning (i.e., training) of the ANN algorithm consists of determining the weights using various methods. One of the popular methods that is based on gradient descent error minimization is known as the backpropagation learning rule [18,19,24]. The backpropagation neural network (BPNN) propagates the error from the output layer to the input layer and improves the accuracy by changing the weights and biases. The weights are typically adjusted after each training epoch to minimize the error between the target and predicted output. The error function of the network is given in the following equation:

$$E = \frac{1}{2}\sum_{i=1}^{n} (T_i - P_i)^2 \qquad (5.7)$$

where T_i is the target for ith input pattern and P_i is the predicted output from the network for the same input pattern.

Multilayer perceptron (MLP) NN is implemented in MATLAB and the main MATLAB functions that we might consider are:

- Feedforwardnet, which creates the network architecture.
- Configure, which sets up the parameters of the network.
- Train, which trains the network.
- Sim, which simulates the network, by computing the outputs for a given test data.
- Perform, which computes the performance of the network test data whose labels are known.

5.4.4 SUPPORT VECTOR MACHINE

Support vector machines (SVMs) are risk-based supervised learning methods that classify data patterns by identifying a boundary with the maximum margin between data points of each class [25]. The general idea is to find a function with a set error margin that maps input variables to output variables such that the predicted output does not deviate from the actual output more than the set error margin (Figure 5.7).

Given the set of labeled data points (input_output pairs) $S = \left\{(x_i, y_i)\right\}_{i=1}^{n}$ where $y_i \in \{-1, +1\}$ is the class label of a high-dimensional point x_i, that is, $(x_i, y_i) \in R^{d+1}$; and d and n are the numbers of features and labeled data points, respectively. In a binary classification problem, points labeled with +1, and −1 belong to classes C^+, and C^-, respectively, that is, $S = C^+ \cup C^-$. The optimal classifier (also known as soft margin SVM) is determined by the parameters w and b through solving the convex optimization problem:

$$\text{minimize } \frac{1}{2} w^2 + C \sum_{i=1}^{n} \zeta_i \tag{5.8}$$

$$\text{Subject to } y_i \left(w^T x_i + b \right) \geq 1 - \zeta_i, \, i = 1, \ldots\ldots, n \tag{5.9}$$

$$\zeta_i \geq 0, \, i = 1, \ldots\ldots, n \tag{5.10}$$

with corresponding linear prediction function $f(x) = wx + b$.

The magnitude of penalization for misclassification is controlled by the parameter C and slack variables ζ_i.

In various difficult cases (that are often observed in practice), the data is not separable into two classes by a hyperplane determined by w. A common approach to cope with this problem is to map S to high-dimensional space and find a hyperplane there. Many existing implementations of SVM solve the Lagrangian dual problem [26] instead of the primal formulation above due to its fast and reliable convergence.

When using SVM, one has to be very careful about the types of data, algorithms, and implementation. An imbalanced data (the sizes of C^+, and C^- can substantially differ from each other) is one of the frequently observed potential problems of SVM models. Correct classification of small class points often becomes more important than classification of others (e.g., in the healthcare domain, the number of sick people is less than healthy). In such cases, one should look for a special SVM model that addresses this issue, for example, weighted SVM that introduces different misclassification penalties for different classes. Determining the correct type of mapping to high-dimensional space can also be extremely important.

Training a typical SVM model becomes time-consuming when the number of points or dimensionality is big. Solvers for Lagrangian dual problem typically scale between $O(dn^2)$ to $O(dn^3)$. In order to train classifiers for large-scale data sets, one should consider using strategies such as parallel implementations [27], memory-efficient solvers [28], and hierarchical approaches [29]. In MATLAB, SVM is implemented with the function $\text{fitcsvm}(X; y)$, where X is the matrix of the training observations and y is a vector of the training labels. We can specify the kernel function, kernel scale parameter, initial estimates of Lagrange multipliers, and other parameters.

REFERENCES

1. Ross, T. (1938). The synthesis of intelligence-its implications. *Psychological Review*, 45(2), 185.
2. Samuel, A. L. (1959). Some studies in machine learning using the game of checkers. *IBM Journal of Research and Development*, 3(3), 210–229.

3. Zhang, J. Wang, F.-Y., Wang, K., Lin, W.-H., Xu, X., & Chen, C. (2011). Data-driven intelligent transportation systems: a survey. *IEEE Transactions on Intelligent Transportation Systems*, 12(4), 1624–1639.
4. Agrawal, R., Imieli´nski, T., & Swami, A. (1993). Mining association rules between sets of items in large databases. In *ACM SIGMOD Record, No. 22*, ACM, New York, NY (pp. 207–216).
5. Meyer, M., & Miller, E. (2001). Urban transportation planning: a decision-oriented approach. *Journal of Transportation Engineering*, 127, 1–642.
6. Tarassenko, L. (1998). *Guide to Neural Computing Applications*. Oxford, UK, Butterworth-Heinemann.
7. Hasson, F., Keeney, S., & McKenna, H. (2000). Research guidelines for the Delphi survey technique. *Journal of Advanced Nursing*, 32(4), 1008–1015.
8. Jain, A. K., Duin, R. P. W., & Mao, J. (2000). Statistical pattern recognition: a review. *IEEE Transactions on Pattern Analysis and Machine Intelligence*, 22(1), 4–37.
9. Polikar, R. (2006). Ensemble based systems in decision making. *IEEE Circuits and Systems Magazine*, 6(3), 21–45.
10. Durrant-Whyte, H. F. (1988). Sensor models and multisensor integration. *International Journal of Robotics Research*, 7(6), 97–113.
11. Castanedo, F. (2013). A review of data fusion techniques. *Scientific World Journal*, 2013, 704504.
12. Heath, M. T. (2002). *Scientific Computing*. New York, McGraw-Hill.
13. Marczak, F., & Buisson, C. (2012). New filtering method for trajectory measurement errors and its comparison with existing methods. *Transportation Research Record: Journal of the Transportation Research Board* 2315, 35–46.
14. Rousseeuw, P.J., & Leroy, A. M. (2005). *Robust Regression and Outlier Detection*. New York, John Wiley & Sons.
15. Friedman, J., Hastie, T., & Tibshirani, R. (2001). *The Elements of Statistical Learning*. Berlin, Springer.
16. Patel, J. A., & Sharma, S. (2014). Big data for better health planning. *International Conference on Advances in Engineering & Technology Research (ICAETR – 2014)*, Unnao, India, 2014, pp. 1–5, https://doi.org/10.1109/ICAETR.2014.7012828.
17. Wu, X., Kumar, V., Quinlan, J. R., Ghosh, J., Yang, Q., Motoda, H., et al. (2008). Top 10 algorithms in data mining. *Knowledge and Information Systems*, 14(1), 1–37.
18. Duda, R. O., Hart, P. E., Stork, D. G. (2012). *Pattern Classification*. New York, John Wiley & Sons.
19. Alpaydin, E. (2014). *Introduction to Machine Learning*. MIT Press.
20. Breiman, L., Friedman, J., Stone, C. J., & Olshen, R. A. (1984). *Classification and Regression Trees*. CRC Press. https://doi.org/10.1201/9781315139470
21. McCulloch, W. S., & Pitts, W. (1943). A logical calculus of the ideas immanent in nervous activity. *Bulletin of Mathematical Biophysics*, 5(4), 115–133.
22. Dougherty, M. (1995). A review of neural networks applied to transport. *Transportation Research Part C: Emerging Technologies*, 3(4), 247–260.
23. Karlaftis, M. G., & Vlahogianni, E. I. (2011). Statistical methods versus neural networks in transportation research: differences, similarities and some insights. Transportation Research Part C: Emerging Technologies, 19(3), 387–399.
24. Werbos, P. J. (1990). Backpropagation through time: what it does and how to do it. *Proceedings of the IEEE*, 78(10), 1550–1560.
25. Vapnik, V. (2013). *The Nature of Statistical Learning Theory*. New York, Springer Science & Business Media.
26. Fletcher, R. (2013). *Practical Methods of Optimization*. New York, John Wiley & Sons.

27. Zhu, K., Wang, H., Bai, H., Li, J., Qiu, Z., Cui, H., et al. (2008). Parallelizing support vector machines on distributed computers. *Advances in Neural Information Processing Systems* 20, Proceedings of the Twenty-First Annual Conference on Neural Information Processing Systems, 257–264, Vancouver, British Columbia, Canada, December 3–6, 2007.
28. Chang, C.-C., & Lin, C.-J. (2011). LIBSVM: a library for support vector machines. *ACM Transactions on Intelligent Systems and Technology*, 2(3), 27.
29. Razzaghi, T., & Safro, I. (2015). Scalable multilevel support vector machines. *Procedia Computer Science*, 51, 2683–2687.

6 ITS and Sustainability

6.1 SUSTAINABILITY

Most organizations today strive to achieve sustainability. Although there are many definitions given to sustainability, there is no agreed definition. The idea of sustainability came from the world's first Earth Summit in Brundtland's [1] Report for the World Commission on Environment and Development [1] defined sustainable development as "Development that meets the needs of the present without compromising the ability of future generations to meet their own needs."

Sustainability is a process or state that can be maintained at a certain level for as long as is wanted. As a noun, it means the ability to be sustained, supported, upheld, or confirmed. In environmental sciences, the term "sustainable" has been employed for processes that do not harm the environment, do not exhaust all natural resources, and thereby support an ecological balance for a long-term period.

6.1.1 Why Sustainability Is Important

The aim of sustainability is to improve the environmental health and quality of life for our society. There are many reasons why sustainability is important for our lives. We need clean air, natural resources, and a nontoxic environment in order to maintain a healthy community. Sustainability must be guaranteed not only for the current generation but also across generations, not only for a country or an organization but also for the wider global community.

6.1.2 The Pillars of Sustainability

Sustainability can be grouped into three categories, or "pillars": environmental, social, and economic. The three pillars must be supported. Sustainability cannot work if one of the pillars is not supported. These three pillars are informally referred to as people, planet, and profits.

6.1.2.1 The Environmental Pillar

The environmental pillar is the commonest one. Its aim is to reduce the impact of human activities on the natural systems that support the community and environment. These involve reducing carbon footprints, packaging waste, water usage, and their overall effect on the environment. Doing so will have a beneficial impact on the planet and can also have a positive financial impact.

6.1.2.2 The Social Pillar

Social sustainability is to develop processes, that maintain a healthy community for its current members but also for future generations. It must have the support and

DOI: 10.1201/9781032691787-6

approval of its employees, stakeholders, and the community where it operates. This involves treating employees fairly and being a good neighbor and community member, both locally and globally. A business should be aware of how its supply chain operates, whether child labor is used, and whether are workers being paid fairly?

6.1.2.3 The Economic Pillar

A sustainable business must make a profit. The other two pillars such as the environment and social pillars cannot be compromised simply to make a profit. However, profit cannot compromise the other two pillars. In order to balance between the three pillars, the economic pillar should include compliance with standards, proper business governance, and risk management. A balance between the following goals is usually attempted: keep a healthy and equitable society, protect the environment, and guarantee economic prosperity. The three goals are interrelated and equally contribute to the economic sustainability of the development plan so this balance throughout the plan lifecycle is important.

6.2 SUSTAINABLE TRANSPORTATION

There is growing interest in sustainable transport. The subject has a huge impact on the sustainability of the planet. It is generally agreed that current levels of car use, fuel consumption, and emissions are unsustainable [2] (See Figure 6.1).

Sustainable transportation is a growing concern in urban areas because of increasing urban populations and the recognition of urban contributions to climate change. Cities are now home to a large proportion of the human population and are growing every year. In order to ensure urban survival and productivity, cities need to provide affordable, accessible, environment-friendly transportation systems. Additionally, in the face of climate change, urban mobility needs to be addressed more sustainably.

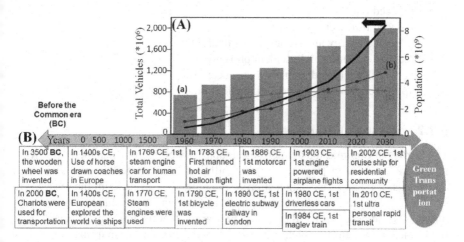

FIGURE 6.1 The growth of total vehicles (black line) against world population growth ((a) total (column), (b) urban (blue line), and (c) rural (red line)); (B) Transportation history.

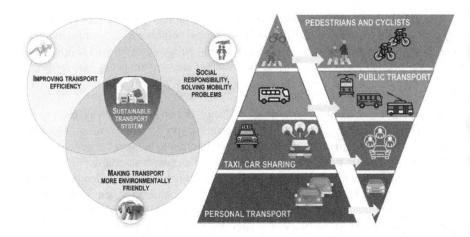

FIGURE 6.2 Key parameters of GT to achieve sustainable mobility in society.

6.2.1 GREEN TRANSPORTATION

Transportation has a significant economic, social, and environmental impact on society (Figure 6.2). Transportation should be given priority to achieve sustainability goals in urban areas; the latter can be achieved by managing the diverse aspects of travel demand (public transportation), vehicle growth patterns, and efficient land-use patterns [3]. Green transportation can be defined as "the transportation service with a fewer negative impact on human health and the environment compared to existing transportation services" [4]. GT can be considered a combinatorial technology comprising the optimal use of traditional fuels, the efficient use of electric vehicle technologies, the use of biogas as a fuel for buses, and strengthened public transportation [5].

An effective GT system can lead to:

- Reduced risk;
- Reduced traffic congestion;
- Enhanced energy and resource sustainability;
- Reduced pollution and accident prevention;
- Increased safety and security assurance;
- Optimized traveling speed and traffic flow.

6.2.2 SUSTAINABLE TRANSPORT

The European Commission defined sustainable transport as one that meets the needs of the present without jeopardizing the ability of future generations to meet their own needs. It refers to the satisfaction of high mobility levels, with the lowest possible energy and environmental costs aimed at satisfying the demand for mobility by businesses and people [6]. Other researchers argue that the term sustainability is "a road transport system in which each user pays at least the full marginal social cost of commuting." The definition Zeitler [7] for "sustainable mobility" employs "... any form of human mobility that responds to the various physical (and social) challenges in the least polluting way ..." and identifies sustainable mobility in the natural way movement, that is, walking and cycling. According to the definition adopted by the Council of Transport Ministers of the European Union, sustainable mobility can be defined as a "transport system and transport model that provides the means and opportunities to meet economic, environmental and social needs efficiently and fairly," while at the same time "it minimizes avoidable or unintended negative effects and their corresponding costs across the different spatial scales."

According to Richardson [8] a sustainable transport system is: "One in which fuel consumption, emissions, safety, congestion, and social and economic access are of such levels that they can be sustained into the indefinite future without causing great or irreparable harm to future generations of people throughout the world". The World Business Council for Sustainable Development defines sustainable mobility as: "the ability to meet society's need to move freely, gain access, communicate, trade, and establish relationships without sacrificing other essential human or ecological values, today or in the future." According to Litman and Burwell [9], the aim of sustainable transportation is to ensure that environmental, social, and economic considerations are involved in decisions affecting transportation activity. Sustainable transport systems must provide a basic requirement to meet society's and the economy's mobility needs as well as social equity [10].

6.2.3 WHY SUSTAINABLE TRANSPORT IS IMPORTANT

Currently, 54% of the world's population lives in an urban area and the United Nations predicts that this number will rise to 66% by the year 2050. Traffic congestion is a daily occurrence for commuters in urban areas. This contributes to climate change by producing volumes of greenhouse gas (GHG) emissions [11]. GHG emissions are identified as a contributor to the changing climate [11]. According to Rice [12] and EPA [13], climate change will increase overall variability of temperature, precipitation, and wind patterns. This variability will increase the probability of heat waves, flooding, and/or drought. Thus traffic congestion problems affect society's quality of life socially, economically, and environmentally. Intelligent transport systems (ITS) have the potential to alleviate the unsustainable impacts of traffic congestion and to improve sustainable transportation systems in the world.

According to Sim et al. [14], the most important factor for air pollution in the urban environment is road traffic. In 1990, traffic accident deaths ranked ninth among the leading causes of death worldwide, with an estimated 2020 rise to the third position. Estimates of the economic cost of these losses amount from 2% to 4% of GDP in developed countries. According to several reports, 65% of the European population is systematically exposed to noise levels above 55 dB, high enough to cause nuisance, aggression, and sleep disturbance [15]. In economic terms, congestion is estimated to cost around 2% of GDP in the European Union, with air pollution and noise at least 0.6% of GDP.

In most European countries, the problem of transport is quite acute, with the continuing increase in demand for both passengers and freight, and this is directly linked to the economic development of the European Union. However, this demand is not accompanied by a corresponding increase in the available transport infrastructure and the means of transport available. Thus sustainable transport systems are critical both for quality of life and for the environment at a global and a local level. This means reducing the impact of transport on the economy, society, and the environment.

6.3 INTELLIGENT TRANSPORT SYSTEMS

ITS combines the implementation of technological improvements to a road system with improvements that increase the road system's efficiency. The goals of ITS are to enhance mobility, increase fuel efficiency, accessibility, operating efficiency, safety, and reduce pollution. According to Bekiaris and Nakanishi [16], ITS helps mitigate problems such as traffic congestion, air quality, and safety without constructing additional roads. ITS includes advanced traffic management systems, advanced traveler information systems, advanced public transportation systems, commercial vehicle operations, and others. Traffic congestion has economic, environmental, and safety impacts on society. ITS can provide benefits to reduce traffic congestion and improve road system efficiency and safety [17].

According to Haque et al. [18], three key elements of modern transportation are sustainability, safety, and smartness. Because of increasing concerns about environmental issues and climate change, the sustainability of a transportation system has become very important. Besides environmental issues, other intractable problems of a land transport system include traffic fatalities and injuries, congestion, noise pollution, depletion of resources, and inaccessibility to facilities. Traffic accidents and congestion impose a huge economic burden on society. Many strategies and policy directions have been developed over the years to address the problems. According to Haque et al. [18], these include: integrating land use and transport planning [14], designing compact-city plans [19], implementing transit-oriented developments [20], controlling the growth of motorization [21], managing travel demand through pricing and financing [22], promoting public transport [23], increasing walking and cycling facilities [24], and incorporating environment-friendly technologies [25]. These strategies and policies are indeed helpful to create a sustainable transportation system that seeks a proper balance between transportation needs and available resources within and between current and future generations. The author agrees with Haque et al. [18] that a clear vision of a transport system is needed for identifying and developing appropriate strategies and policies to build an efficient, long-lasting, and safe transport system.

6.3.1 IMPLEMENTING SUSTAINABLE ITS

Because of its social, technical, and economic components, implementing a sustainable transport system is a complex process. According to Haque et al. [18], the planning of transport systems is usually framed by articulating principles and desirable attributes. May et al. [26] suggested that there are six overarching objectives of a sustainable transport system. These are:

1. Economic efficiency,
2. Livable streets and neighborhoods,

3. Protection of the environment,
4. Equity and social inclusion,
5. Health and safety, and
6. Contribution to economic growth.

Castillo and Pitfield [27] using these objectives have developed an evaluative and logical approach to identify and rank sustainable transport indicators based on measurability, availability, and interpretability. In their paper, these authors have several indicators showing that managing traffic volume, encouraging cycling trips, promoting public transport, reducing CO_2 and other air pollutant emissions, and lowering traffic accidents were the key indicators of sustainability.

According to ECMT [28], a sustainable transport system is one that:

1. allows for the safe and environmentally harmless basic means of access and development on the individual, business, and societal level, while promoting equity within and between generations;
2. is reasonably priced and runs efficiently, providing choice of transport mode as well as support for a competitive economy and good regional development;
3. Keeps production of emissions and waste within the carrying capacity of the natural environment and keeps the consumption of renewable resources and nonrenewable resources respectively within the rates of generation and development of renewable substitutes, while minimizing the impact on the use of land as well as production of noise. This has three major targets:
   1. Economic development,
   2. Environment protection, and
   3. Social equity.

Haque et al. [18] argue that smart technologies can be used to promote sustainability. Their work in the Singapore experience shows that smart technologies help to implement or escalate various policies and strategies related to sustainability. They mentioned that smart technologies like bus priority signal systems, bus lane enforcement systems, availability of real-time service information, and an integrated multimodal fare payment technology have been helpful to promote public transport as a viable alternative to private transport. Traffic signal coordination system using GLIDE helps to ensure a smooth flow along the corridors and hence reduces congestion, fuel consumption, and emissions. Smart taxi booking systems and public transport information-sharing systems have increased the accessibility for commuters. In addition, the availability of real-time traffic and travel-related information has enhanced motorists' flexibility in route planning for a less congested, faster, and safer trip. The electronic toll payment system is another smart technology, that has been successfully implemented to facilitate the road pricing policy for managing congestion and hence promoting sustainability.

A transport system consists of the means, equipment, and logistics required for the carriage of passengers and goods, including transport networks and infrastructures, nodes, and connections, modes of transport as well as policies for their smooth operation. In addition, the purpose of the transportation system is to coordinate the movement of persons, goods, and vehicles. On the other hand, mobility as a term refers to the ability of people and goods to move easily, quickly, and economically, where it is intended at a speed that reflects free flow or comparatively high-quality conditions [29]. In other words, mobility expresses the ability to arrive at the destination in a time and cost-efficient manner [30]. The definition of a sustainable transport system differs from sustainable mobility in that one does not exclude the other, in many ways overlapping or complementing each other [31].

Priority must be given to the individual parts of the transport system in order to achieve sustainable mobility. The choice of the quality of transport systems for vehicles (and therefore drivers) or pedestrians should be made. For example, speed or traffic safety, public transport, or private vehicles will be given priority [32]. In the last 10 years, the European Commission has often referred to urban sustainable mobility as part of the goals and actions to achieve sustainable urban development. Specific guidelines have been proposed for the implementation of sustainable transport systems, such as improving public transport and encouraging mild forms of transport (pedestrian, bicycle). As a result, member states have been adopting urban mobility action plans. The proposed measures in the context of the basic guidelines for urban mobility are:

1. Unified Spatial Planning, Urban Planning, and Transport to address major cities' problems in the field of transport by implementing policies aimed at improving traffic, and giving emphasis on public transport infrastructure.
2. Traffic Management with the promotion of public transport, instead of private vehicle traffic (priority to means of mass transport, smart traffic, integrated parking policy, road safety upgrade, etc.).
3. Mild Refurbishments that give pedestrians and cyclists significant usable space.
4. Technologies and Measures for the Environment, such as Vehicle and Fuel Emission Reduction Technologies and Environmental Pricing based on the "polluter pays" principle.

In the context of achieving sustainable mobility, other programs have been developed at the regional (e.g., European) level. One of them is the ATTAC project (Attractive Urban Public Transport for Accessible Cities), which aims to improve coordination in the promotion, design, and operation of public transport networks.

6.3.2 Sustainable Urban Mobility Plans

The European Commission, in the Urban Mobility Action Plan, has decided to speed up the uptake of Sustainable Urban Mobility Plans (SUMPs). SUMPs are essentially strategic urban mobility action plans, that aim to create a sustainable urban transport system based on existing planning practices, covering the mobility needs of individuals today and in the future. The White Paper on Transport [33], it proposed to examine the feasibility of SUMPs as a mandatory approach for cities of a certain size, as well as the possibility of creating a European support framework for the implementation of SUMPs in European cities (SUMPs, Planning for People).

The EU countries that are pioneering the adoption of SUMP principles are the United Kingdom (with Local Transport Plans—LTP) and France (with Plande Déplacements Urbains—PDU).

Overall, the objectives of SUMPs are:

a. to establish urban transport and sustainable mobility;
b. to ensure the accessibility of jobs and services to everyone;
c. to improve security and safety;
d. to reduce pollution, energy consumption, and gas emissions;
e. to increase efficiency and cost-effectiveness for the transport of people and goods;
f. to enhance the quality and appeal of the urban environment;
g. to improve the health of residents and the quality of the urban environment;
h. to improved accessibility and mobility;
i. to promote of public transport;
j. to develop better spatial planning plans, etc.;
k. to provide a better quality of life;
l. to use available resources effectively.

6.4   WHY WE NEED TO HAVE SUSTAINABLE ITS

Transport plays a key role in generating economic progress through trade and mobility, but it also accounts for global CO_2 emissions and air pollution worldwide, particularly due to road freight transport. In this sense, these sectors urgently need to be transformed and made more sustainable. The author concurs with Stephenson, Spector et al. [34] that unsustainable consequences include environmental impacts, for example [35]; social impacts, for example [36]; and economic impacts for example [37].

It is imperative that we move to more sustainable transport systems. According to Stephenson et al. [34], this requires market-based solutions such as shared mobility businesses and the increasing cost-competitiveness of electric vehicles. This is unlikely to occur without carefully designed and integrated government interventions [38].

6.5   SUSTAINABLE DEVELOPMENT GOALS FOR ITS

Sustainable development is a complex concept that is subject to numerous interpretations because it involves several disciplines and possible interconnections. It must satisfy the needs of the present without compromising the capacity of future generations, guaranteeing the balance between economic growth, care for the environment, and social well-being.

On January 1, 2016, the 2030 Agenda for Sustainable Development adopted by world leaders in September 2015, officially came into force. The 2030 Agenda is a set of 17 sustainable development goals (SDGs) with 169 targets stimulating actions to move the world onto a sustainable and resilient path. It is an important roadmap guiding policy actions for sustainable development in the next 15 years. The sustainable development must be able to protect the planet and guarantee the global well-being of people. These goals require the active involvement of individuals, businesses, administrations, and countries around the world. They are known as the global goals. These goals are a call from the United Nations to all countries around the world to address the great challenges that humanity faces and to ensure that all people have the same opportunities to live a better life without compromising our planet.

The 2030 Agenda states that sustainable transport systems, along with universal access to affordable, reliable, sustainable, and modern energy services, quality and resilient infrastructure, and other policies that increase productive capacities, would build strong economic foundations for all countries. The text includes five targets that are directly related to the transport sector and seven other targets that are indirectly related to the transport sector.

Transport contributes directly to five targets on road safety (Target 1); energy efficiency (Target 2); sustainable infrastructure (Target 3), urban access (Target 4), and fossil fuel subsidies (Target 5) emphasize that sustainable transport is not needed solely for its own sake, but rather is essential to facilitate the achievement of a wide variety of SDGs.

Transport also contributes indirectly to seven SDG targets on agricultural productivity (Target 1), air pollution (Target 2), access to safe drinking water (Target 3), sustainable cities (Target 4), reduction of food loss (Target 5), climate change adaptation (Target 6), and climate change mitigation (Target 7).

Sustainable transportation is the capacity to support the mobility needs of a society in a manner that is the least damageable to the environment and does not impair the mobility needs of future generations [39]. It should contribute to environmental, social, and economic objectives. It is therefore important that we have a policy framework to support sustainable transport, which includes low-carbon modes of transport, energy efficiency, user-friendly transport initiatives, integration of transport, and land-use planning.

Besides the three major pillars of sustainable development, Rodrigue et al. [39] argue that we should also consider the question of whether sustainability should be imposed by regulation or be the outcome of market forces. According to Rodrigue et al. [39], societies do not contribute to environmental problems at the same level. These authors believe that sustainability can be expressed on two spatial levels:

- global is long-term stability of the earth's environment and availability of resources to support human activities and
- local that is often related to urban areas in terms of jobs, housing, and environmental pollution.

Although the transport sector has the potential to improve the lives of the people, currently there is a lack of leadership at the global level without a clear set of principles to transform the sector [40], It is important that actions are taken to ensure that we have a plan for sustainable ITS. Possible suggestions are:

- to define clearly the objectives underpinning sustainable mobility;
- to give due consideration for safety, equity, and climate for all road projects;
- to adopt the cocreation of value approach for ITS design;
- to use new technology such as IOT, Big data, and cloud computing to support future mobility;
- to plan a strategy for sustainable transport as a priority at local, national and global levels;
- to promote an integrated approach to policymaking at the national, regional, and local levels for transport services and systems to promote sustainable development;
- to develop alternative means of transport.

To develop sustainable ITS is not trivial. There are many issues that need to be considered. These issues would depend on the economy, social, and priority of the country concerned. No country is an island; the world is interconnected. It is important to remember that there are different stakeholders involved from local, to regional to global levels. Each stakeholder has different values. These different values must be addressed and met. There may be conflicting values between stakeholders. To overcome the different values, we must plan the design of a sustainable ITS based on the co-creation of values.

6.6 TRANSPORT POLICY DEVELOPMENT

Three important legislative documents in the field of transport have emerged since the current government came into power. In [33], followed by the National Land Transport Transition Act [41].

The aim of these documents is twofold. Firstly, it provides a framework of procedures indicating which documents, permits, etc. are required by the national government [41]. Municipalities and operators also have to comply and provide the required documents timeously.

The second, more important, aim is to provide a policy focus. This focus can be summarized as the aim to provide mobility for all. The vision statement is as follows [42]: "Provide safe, reliable, effective, efficient, and fully integrated transport operations and infrastructure which will best meet the needs of freight and passenger customers at improving levels of service and cost in a fashion which supports government strategies for economic and social development while being environmentally and economically sustainable."

The strategies for implementation are the promotion of integration and intermodalism. In other words, promotion of public transport is key.

The administration analyzed the current market and identified challenges. With regards to urban passenger customers, the main focus of this dissertation, different segments with specific needs were identified, and specific issues were generated in respect of each segment. A large number of "Stranded" passengers were identified and predicted to grow well to over 3 million. The core challenges that emerged were [43]:

- The lack of affordable basic access;
- The ineffectiveness of the public system for commuters and other users;

- The increasing dependency on cars within the system; and
- The impact of past land-use patterns and existing planning and regulation of public transport.

Based on the customer analysis, three key thrusts have been adopted for urban areas [17]:

1. High-volume corridors: Densification in corridors and nodes to achieve economies of scope, effectively turning around the current trend toward dispersal;
2. Sustainable operations: Optimize modal economics and service mix through infrastructure investment to support the corridors, and by selecting the optimal mode based on the cost/service trade-off. This involves also facilitating differentiated service and choice, wherever possible, but with subsidization only for the optimal mode, if at all. Tough road-space management is necessary to prioritize public transport and subsidies should be targeted at affordable access to the optimal mode, and
3. Improved efficiency: Improve firm-level performance through competitive tendering to the private sector with incentives for productivity innovations, effectively regulating all modes, especially taxis, and improving sustainability through capital investment.

REFERENCES

1. Brundtland, G. H. (1987). Report of the World Commission on environment and development: our common future. United Nations.
2. Goldman, T., & Gorham, R. (2006). Sustainable urban transport: four innovative directions. *Technology in Society*, 28(1–2), 261–273.
3. Sangaradasse, P., & Eswari, S. (2019). Importance of traffic and transportation plan in the context of land use planning for cities - a review. *International Journal of Applied Engineering Research*, 14, 2275–2281. www.ripublication.com/ijaer19/ijaerv14n9_33. pdf. (Accessed 12 February 2024).
4. Bjorklund, M. (2011). Influence from the business environment on environmental purchasing - drivers and hinders of purchasing green transportation services. *Journal of Purchasing & Supply Management*, 17, 11–22. https://doi.org/10.1016/j. pursup.2010.04.002.
5. Lee, C. T., Hashim, H., Ho, C. S., Fan, Y. V., & Klemes, J. J. (2017). Sustaining the low-carbon emission development in Asia and beyond: sustainable energy, water, transportation and low-carbon emission technology. *Journal of Cleaner Production*, 146, 1–13.
6. Dirks, S., & Keeling, M. (2009). A vision of smarter cities: how cities can lead the way into a prosperous and sustainable future. IBM Institute for Business Value, 8. http://www-935.ibm.com/services/us/gbs/bus/pdf/ibm_podcast_smarter_cities.pdf
7. Zeitler, U. (2015). *Environmental Ethics and Sustainable Mobility. Centre for Social Science Research on the Environment (CESAM)*. Aarhus University, Denmark.
8. Richardson, B. (1999). Towards a policy on a sustainable transportation system. *Transportation Research Record*, 1670, 27–34.
9. Litman, T., & Burwell, D. (2006). Issues in sustainable transportation. *International Journal of Global Environmental Issues*, 6(4), 331–347.

10. Williams, K. (2005). Spatial planning, urban form and sustainable transport: an introduction. In K. Williams (Ed.), *Spatial Planning, Urban Form and Sustainable Transport* (pp. 1–13). Aldershot, Burlington, Ashgate.
11. United Nations Habitat. (2015). Climate Change. Available from: https://unhabitat.org/ urban-themes/climate-change/ Accessed 01.03.15.
12. Rice, D. (2014). Cities in U.S. Will Absorb the Heat of Climate Change. Available from: https://www.usatoday.com/ story/weather/2014/05/07/climate-changeimpacts/8810849/ Accessed 11.04.15.
13. EPA. (2014). Impacts & Adaptation. Available from: https://www.epa.gov/climate-change/impacts-adaptation/ Accessed 11.04.15.
14. Sim, L. L., Malone-Lee, L. C., & Chin, K. H. L. (2001). Integrating land use and transport planning to reduce work-related travel: a case study of Tampines regional centre in Singapore. *Habitat International, 25*(3), 399–414.
15. Abo-Qudais, S, & Abu-Qdais, H., (2005). Perceptions and attitudes of individuals exposed to traffic noise in working places. *Building and Environment*, 40(6), 778–787.
16. Bekiaris, E., & Nakanishi, Y. (2004) *Economic Impacts of Intelligent Transportation Systems: Innovations and Case Studies*. Amsterdam, The Netherlands, Elsevier. Available from: https:// medcontent.metapress.com/index/A65RM03P4874243N.pdf.
17. World Health Organization (WHO). (2015). *Global Status Report on Road Safety*. WHO Library Cataloguing-in-Publication Data. UNESDOC Digital Library.
18. Haque, M. M., Chin, H. C., & Debnath, A. K. (2013). Sustainable, safe, smart-three key elements of Singapore's evolving transport policies. *Transport Policy*, 27, 20–31.
19. Sung, H., & Choo, S. (2010) Policy implications of compact-city measures for sustainable development: a case study in Korea. In *Proc. 89th Annual Meeting of Transportation Research Board*, Washington, DC, USA.
20. Sung, H., & Oh, J.-T. (2011) Transit-oriented development in a high-density city: identifying its association with transit ridership in Seoul, Korea. *Cities*, 28(1), 70–82.
21. Han, S. S. (2010). Managing motorization in sustainable transport planning: the Singapore experience. *Journal of Transport Geography*, 18, 314–321.
22. Seik, F. T. (2000). An advanced demand management instrument in urban transport: electronic road pricing in Singapore. *Cities*, 17(1), 33–45.
23. Ibrahim, M. F. (2003). Improvements and integration of a public transport system: the case of Singapore. *Cities*, 20, 205–216.
24. Duduta, N., Shirgaokar, M., Deakin, E., & Zhang, X. L. (2010). An integrated approach to sustainable transportation, land use and building design; the case of the Luokou District, Jinan, China. In *Proc. 89th Annual Meeting of Transportation Research Board*, Washington, DC, USA.
25. Lopez-Ruiz, H. G., & Crozet, Y. (2010) Sustainable transport in France: Is a 75% reduction in CO_2 emissions attainable? In *Proc. 89th Annual Meeting of Transportation Research Board, 2010*. Washington, DC, USA.
26. May, A.D., Jarvi-Nykanen, T., Minken, H., Ramjerdi, F., Matthews, B., & Monzon, A. (2001). *Cities' Decision-Making Requirements - PROSPECTS Deliverable 1*. Leeds, Institute of Transport Studies, University of Leeds.
27. Castillo, H., & Pitfield, D. E. (2010). Elastic - a methodological framework for identifying and selecting sustainable transport indicators. *Transportation Research Part D: Transport and Environment*, 15(4), 179–188.
28. ECMT. (2001) European conference of ministers of transport: transport/telecommunications. In *2340th Council Meeting, 7587/01 (Presse 131)*, Luxembourg, 4–5 April 2001. Available from: https://corporate. skynet.be/sustainablefreight/trans-counci-conclusion-05-04-01.htm Accessed 24.06.10.
29. Paris Declaration. (2015). *City in Motion: People First*. Geneva, United Nations Publications.

30. Hollands, R. G. (2008). Will the real smart city please stand up? Intelligent, progressive or entrepreneurial? *City*, 12, 303–320.
31. Bifulco, F., Amitrano, C. C., & Tregua, M. (2014). Driving smartization through intelligent transport. *Chinese Business Review*, 13(4), 243–249
32. Hernández-Muñoz, J. M., Vercher, J. B., Muñoz, L., Galache, J. A., Presser, M., Gómez, L. A. H., & Pettersson, J. (2011). Smart cities at the forefront of the future internet. In Domingue, J. et al. (eds.), *The Future Internet Assembly* FIA 2011. Lecture Notes in Computer Science, vol. 6656 (pp. 447–462). Berlin, Heidelberg, Springer. https://doi.org/10.1007/978-3-642-20898-0_32
33. European Commission. (2011). Directorate-general for mobility and transport. White Paper on Transport: Roadmap to a Single European Transport Area: Towards a Competitive and Resource-efficient Transport System. Publications Office of the European Union.
34. Stephenson, J., Spector, S., Hopkins, D., & McCarthy, A. (2017). Deep interventions for a sustainable transport future. *Transportation Research Part D: Transport and Environment*. https://doi.org/10.1016/j.trd.2017.06.031.
35. Hopkins, D., & Higham, J. E. S. (2016). *Low Carbon Mobility Transitions*. Oxford, UK, Goodfellow Publishers.
36. Lucas, K. (2012). Transport and social exclusion: where are we now? *Transport Policy*, 20, 105–113.
37. Wallis, I. P., & Lupton, D. R. (2013) *The Costs of Congestion Reappraised*. New Zealand Transport Agency. (No. 489). Wellington, New Zealand.
38. Geerlings, H., Shiftan, Y., & Stead, D. (Eds.). (2012). *Transition towards Sustainable Mobility: The Role of Instruments, Individuals and Institutions*. Surrey, UK, Ashgate Publishing Ltd.
39. Rodrigue, J.-P., Comtois, C., & Slack, B. (2017). *The Geography of Transport Systems* (3rd ed.). New York, Routledge, 440 p.
40. Mohieldin, M., & Vandycke, N. (2017) Sustainable Mobility for the 21st Century. Available from: https://www.worldbank.org/en/news/feature/2017/07/10/sustainable-mobility-for-the-21st-century.
41. National Department of Transport (NDoT). (2000). National Land Transport Transition Act (NLTTA), Pretoria, SA.
42. National Department of Transport (NDoT). (1996). White Paper on National Transport Policy, Pretoria, SA.
43. National Department of Transport (NDoT). (1999). Moving South Africa: A Transport Strategy for 2020, Pretoria, SA.

7 Moving Toward ITS via Internet-of-Things (IoT)

7.1 BACKGROUND

To build an Intelligent Transportation System (ITS) using the Internet of Things (IoT) platform, the system has three components: the sensor system, the monitoring system, and the display system [1]. The sensor system has Global Positioning System (GPS), Near Field Communication (NFC), Temperature and Humidity sensors, which are always connected to the Internet via a GSM network to track the location, commuter, and ambience inside the bus. The monitoring system is used to extract the raw data from the sensors database, convert it into a meaningful context, trigger some events within the bus, and provide information to the bus driver. The display system is used to show the context data to all the commuters at the bus stop. Nowadays, there are several use cases related to IoT for ITS, such as connected and autonomous vehicles, cooperative transportation networks, and smart roads in order to improve data propagation and create heterogeneous connectivity and low latency applications in high-capacity environments. ITS techniques can be also applied to logistics, so accuracy on delivery and timing can be improved considering all the involved ecosystems, baseline, and standardized architectures in interconnected Smart Cities for future development and integration IoT for ITS techniques can be also applied to civil structures for big data analytics for intelligent systems [2].

7.2 INTERNET OF THINGS

Internet of Things (IoT) is a new model which refers to the ever-growing network of physical things that carry an Internet Protocol (IP) address to be identified and connected to Internet and provide a set of new services by communicating with other internet-enabled devices and systems. IoT is one of the latest fields due to the increase in the use of Internet in the society which has attracted the interest of many researchers. It has really changed the way of life of society as the things that were just considered dummies have got life due to features such as the capability of being identified, the power of sensing, computing and forming a network. It enables large-scale technological innovations and value-added services that will enable the users to interact with the things that have the capability of being identified and provide the information as per the embedded devices. There are a number of IoT applications that can be classified into different categories such as smart cities, transportation, civil infrastructure, health care systems, retailing, agriculture logistics, and remote monitoring. To focus on the various research issues, there is a need to understand the meaning of IoT and to find out that how it affects daily life as per the new business models. We are still in the initial stages, where the

DOI: 10.1201/9781032691787-7

meaning of IoT is being taken as per individual's vision and requirements [3,4]. This is the reason that there is no universal definition of IoT and we do hear of various terms that may be used for IoT like Web of Things (WoT), Machine to Machine (M2M), Cloud of Things (CoT), etc. Though all these terms are equivalent but as per the opinions of authors these terms reflect the interpretations with respect to their vision.

There are so many IoT applications that integrate physical objects using Internet and transfer the collected data over the network without human assistance. Further, the collected data by the embedded devices in the objects is analyzed to make important decisions that help society to use these applications in the different types of services. The usage of IoT devices has increased the volume of the data and there is a need to make this quantitative data qualitative by using it to make the applications intelligent. So, by applying machine learning (ML) techniques, these applications are becoming intelligent and capable of producing better results by addressing the issues in traffic management and providing safety to the residents. However, still there is a lot of scope to deal with the various problems and apply IoT in different fields to maximize automation. the application of machine learning (ML) techniques on it, is heading towards the realization of smart transportation. The complex interactions among roadways, transportation traffic, environmental elements, and traffic crashes have been explored by various researchers. In this paper, an overview of the various approaches applied using ML techniques has been presented.

The main task of the IoT devices is to collect the data, M2M communication, and in some of the cases they pre-processes the data before sending it to other devices or the cloud. The selection of IoT devices is done by keeping in view the power, cost, and energy consumed while designing. The continuous collection and exchange of data by IoT devices results in Big Data. So, an IoT infrastructure has to implement the handling, storing, and then to analyze the bulk data collected by the devices. The IoT platforms not only store and analyze the data but these are also capable of monitoring and managing the nodes. In addition to this, there are some applications where the IoT devices process the data and which usually happens in the centralized node in cloud computing infrastructure. The data can move to the end networking elements for processing giving a new model called Edge Computing. As these devices are the basic devices, so are not able to handle complex computational tasks. In such cases, a node is required that can upload all the data collected by IoT devices to a central cloud node thus minimizing the overhead. This node should possess adequate resources to accomplish this task. The solution was given by authors [5] by introducing a new term called fog nodes. The fog nodes have the capability to handle big data by providing the storage, computing, and network required for such applications. Further, this data stored in the various cloud nodes can be accessed for advanced analysis and this can be done by using various ML techniques and sharing the same among other devices. It will help in the creation of new smart applications. Such IoT applications for a smart city have been developed which can be grouped into the following categories [3].

7.3 SMART TRANSPORTATION

The most crucial application as there is an urgent need to monitor transportation due to the increase of vehicles in urban areas. Modern vehicles are equipped with various sensors and devices. These devices interact with the devices installed on the roadside of the city and even share vital information through the Internet using mobile devices and this data can be used by the cloud nodes to share crucial instructions with the other vehicles on the road. It is now possible to offer optimized routes to the different categories of vehicles, parking slots, and other useful information to the public that can result in safe driving, energy conversation, and avoid congestion on the roads [6].

7.4   ENVIRONMENTAL CONDITIONS MONITORING

Environmental conditions are monitored by the wireless sensors installed throughout the city and it makes the perfect infrastructure to collect the data required to fulfill the said objective. Many devices can be installed that can sense the pressure, humidity, or wind speed to forecast the weather conditions and help the society to manage the day-to-day work schedules accordingly. Some more smart sensors can also be used to monitor the air and water pollution in the cities. The pollution data record can help the government and non-government organizations to find out the cause of any rise in pollution levels.

7.5   CIVIL INFRASTRUCTURE

Nowadays most Civil Infrastructures are equipped with various sensors, such as accelerometers, and other different types of sensors. These sensors have the capability of transferring data and can communicate with authorized users through the internet. This can provide the users the flexibility of managing these structures irrespective of their physical size. The capabilities of the system can be expanded by changing the infrastructure associated with the Intelligent Transportation System. As the technology in IoT has improved, the modern transportation system has benefitted a lot from it with advanced information technology.

7.5.1   SENSORS FOR CIVIL INFRASTRUCTURES

Structural control and health monitoring as condition monitoring are some essential areas that allow for different system parameters to be designed, supervised, controlled, and evaluated during the system's operation in different processes, such as those used in machinery, structures, and different physical variables in mechanical, chemical, electrical, aeronautical, civil, electronics, mechatronics, and agricultural engineering applications, among others. Continuous monitoring of these structures is a need because these are subject to changes in environmental and operation conditions along their lifetime, which can result in changes and possible failures and damages in all the structure and their components. The proper development of these applications is associated with the use of reliable data from sensors or sensor networks, which requires the use of advanced signal processing techniques, sensor data fusion, and data processing (sometimes in real-time) to produce a reliable system and avoid accidents or failures in the process. In [7], the use of data-driven algorithms is explored at each level of the damage diagnosis as well as the instrumentation and implementation process to show the current state of some of the developments of data-driven SHM. It is necessary to test structures or substructures and materials, in some vulnerable stage of construction or when verifying theoretical calculations. Testing and measuring of certain desired parameters has taken place in the field of civil engineering in the latest century. With recent advancements in Sensor technology, Structural Health Monitoring (SHM) systems have been developed and implemented in various civil structures such as bridges, buildings, tunnels, power plants, and dams [8]. Many advanced types of sensors, from wired to wireless sensors, have been developed to continuously monitor structural conditions through real-time data collection. However, there are still a remarkable number of questions associated with the use of SHM sensors. For designing a SHM system, one of the critical missions is discovering how to determine an appropriate type of sensor that can efficiently meet the scopes of the designed sensing system. This article aims to present a brief review of different types of sensors for structural health monitoring and real-time condition assessment of structures. In addition, the problem of sensor integration and data fusion is addressed. We consider the problem of combining information from diversified sources in a coherent fashion. We assume that at the fusion, the information from various sensors may be available in different forms. For example, data from infrared (IR) sensors may be combined with range radar (RR) data, and further combined with visual images. In each case, the data and information from the different sensors are presented in a different format which may not be directly compatible for all sensors. Furthermore, the available information may be in the form of attributes and not dynamical measurements. A theory for sensor integration and data fusion that accommodates diversified sources of information is presented. Data (or, more generically, information) fusion may proceed at different levels, like the level of dynamics, the level of attributes, and the level of evidence. All different levels are considered and several practical examples of real-world data fusion problems are discussed.

7.5.2 SMART SENSORS FOR STRUCTURAL HEALTH MONITORING

Structural health monitoring heavily relies on collecting accurate and high-quality real-time measurements of structural element conditions, communicating this information with the control system, and signaling necessary warnings should an irregular pattern is ever observed. Sensors for structural health monitoring are designed to facilitate the monitoring process and enable maintenance engineers with decision-making tools, which will ensure the safety of the facility and the public. A typical health monitoring system is composed of a network of sensors being responsible for measuring different parameters relevant to the current state of the structure as well as its surrounding environment, such as stress, strain, vibration, inclination, humidity, and temperature.

The latest advances in research on sensor technology for structural health monitoring have resulted in various types of SHM sensors. The following provides a brief review of the most widely used SHM sensors.

7.5.3 SENSING AND PERCEPTION SYSTEMS FOR ITS

Based on monitoring interests, various types of data about a structure's response may be needed in a SHM program and therefore different types of sensors might be used.

Conventional-used sensors available commercially for a long time include load cells, electrical resistance strain gauges, vibrating wire strain gauges, displacement transducers, accelerometers, anemometers, thermocouples, etc. Instructions on the application of these sensors, protection against mechanical and chemical damage, reduction of noise, and the collection of more representative data are usually available with the sensor datasheets. The spur for the development of novel sensors is driven by the limitations of conventional electric sensing technologies to capture linear or planar continuous information. In instances where numerous points are measured over a huge area, a staggering amount of sensors require separate, individual cables for supplying power and transmitting measurement signals. Therefore, this increases the intricacies of the system and presents a challenge in terms of construction costs, maintenance costs, and operating costs. In this chapter, knowledge of conventional measurements of train, displacement, acceleration, and temperature will be discussed first followed by a detailed discussion about fiber optic sensing technology which has experienced a huge advancement the most recent in bridge SHM. Manufacturers of these sensors and relative DAQ systems are reviewed in Part II of this report.

7.5.3.1 Strain Measurement

When trying to monitor a fatigue crack on a critical component, strain gauges, fiber optic sensors (FOS) or other types of sensing devices can provide rich information about the local behavior within the component. The sensors are mounted directly to the region of interest, presuming access is feasible. The number of sensors, in this case, depends on engineers, who must decide the locations to be monitored on the structure. When access is impossible, the strain in the area of interest can be approximated by measuring the strain in a close neighborhood. The approximation can be based on the numerical model updating using experimental data. The mean square error or the mean absolute error between the actual and the predicted strain level can provide a qualitative description of the quality of the estimation.

7.5.3.2 Foil Strain Gage

Foil strain gauges (Figure 7.1a) have been widely used for strain measurement in experimental stress analysis. Foil strain gauges are generally attached to the surface of structural components and wired to readout units. As the component experiences strain, the change in length at the surface of the component is transmitted to the strain gauge through the connecting substances. From there, the corresponding signal is transmitted to the readout unit through the lead wires. To ensure the output of the readout unit represents the true strain change in the material, it is important to understand the various factors that affect the quality of the measurements [9]. Weldable strain gauges are also available in the market (Figure 7.1b). This kind of gauge consists of a specially manufactured foil strain gauge pre-bonded to a metal carrier for spot welding to steel components. Adhesively bonded strain gauges are preferred over weldable strain gauges when the highest accuracy is desired. However, where bonding conditions are not ideal, the weldable type is the preferred choice. Weldable gauges are more costly than bondable gauges, but the overall installation cost is reduced significantly because of the shorter installation time, and elimination

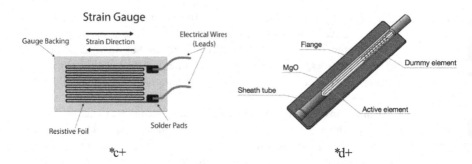

FIGURE 7.1   (a) Foil strain gauges. (b) Weldable strain gauges.

of the strict requirements for surface preparation and adhesive curing required for bondable strain gauges.

Another type of foil strain gauge on the market is the embedment strain gauge. This type of strain gauge is used for measuring strains inside concrete structures. An embedment gauge consists of a long foil gauge (about 100 mm) embedded in a polymer concrete block. The long length of the embedded strain gauge is necessary to ensure that the measured strain is the average strain in aggregate materials and not the localized strain due to discontinuities in concrete. The concrete cover protects the embedment strain gauge against mechanical damage during construction, as well as moisture and corrosive attacks afterward. It also provides a means for the proper transfer of strain from the structure to the strain gauge. The sensitivity of strain gauges to moisture and humidity is another concern, especially when long-term measurement is planned, particularly in a harsh environment, and when it is important to maintain a stable reference (zero-stability) for the gauges. Special provisions are often needed to protect the gauges in order to obtain acceptable measurements. Foil strain gauges are less attractive for field SHM of bridges especially when the distance between the gauge and the readout unit increases. This is due to the fact that the low-level voltage signal produced by the foil strain gauge is susceptible to electromagnetic and electrostatic interference from external sources. When unconditioned signals from foil gauges are transmitted a relatively long distance, the electrical noise superimposed by the electromagnetic and electrostatic fields becomes significant and can lead to inaccurate results and incorrect interpretation of the strain signals. The problem is more severe for dynamic measurements since filtering the noise can change the characteristics of the original signal.

7.5.3.3   Vibrating Wire Sensors

Vibrating wire sensors are a class of sensors that are very popular for geotechnical and structural monitoring purposes. The principal component of the vibrating wire sensor is a tensioned steel wire that vibrates, when pulled, at a resonant frequency that is proportional to the strain in the wire. This mechanism is used to measure static strain, stress, pressure, tilt, and displacement through various sensor configurations. Vibrating wire sensors operate based on the resultant resonance frequency of vibration, rather than amplitude, to convey the signal; hence, these sensors are relatively resistant to signal degradation from electrical noise, long cable runs, and other changes in cable resistance. They have long-term stability and wide usage for monitoring structures such as dams, tunnels, mines, bridges, foundations, piles, unstable slopes, and excavations.

7.5.3.4 Vibrating Wire Strain Gauges

Vibrating wire (VW) strain gauges are encased in sealed steel tubes. The gauges are equipped with a magnet/coil assembly for exciting the wire inside the tube, and sensing its frequency. In some of the models available on the market, the magnet/coil assembly is attached to the outside of the tube, and in others, it is built inside the tube. In general, VW gauges are not susceptible to humidity, however, the surface gauges should be protected against direct contact with the weather. A temperature sensor is a standard feature on every VW strain gauge. The same readout unit used to read the strain reads the temperature sensor. The temperature reading lets the user apply the necessary correction for the temperature effect. Since the temperature-induced strain is the result of the difference in thermal expansion coefficients of the gauge and the instrumented component, no temperature correction is considered when the gauge is attached to a steel component. Vibrating wire (VW) strain gauges are relatively bulky (usually larger than 100 mm in length) and are produced for embedment in concrete or attaching to the surface of components. Surface strain gauges can be welded, bolted, or bonded to the material. Embeddable strain gauges can be directly placed in concrete or cast into a concrete briquette before being placed in their final position. In either case, the placement of large-diameter aggregates in the proximity of the gauge must be avoided. This is essential for preventing stress discontinuity in the gauge area. Theoretically, the maximum aggregate size, within an envelope of 1.5 gauge lengths around the gauge, should not exceed 1/5 of the gauge length.

7.5.3.5 Vibrating Wire Strain Gauge

Vibrating wire strain gauges are widely used to measure strain in steel or reinforced concrete. They can be easily embedded in concrete for monitoring strain in piles, foundations, dams, tunnels, etc. Arc-weldable gauges are suitable for arc welding to steel structures such as tunnel linings, piles, and bridges.

7.5.3.6 Vibrating Wire Displacement Transducer

Vibrating wire displacement transducers are basically designed to monitor long-term movement in critical structures. These sensors are able to measure minor displacements across joints and cracks in concrete, rock, soil, and structural members. In essence, the transducer is composed of a vibrating wire connected to a tension spring. Any displacements of structural in the vicinity of the sensor are accommodated by a stretching of the tension spring, which in turn produces a commensurate increase in wire tension. These sensors are mostly used for crack width measures for example in bridges and tunnels.

7.5.3.7 Linear Variable Differential Transducers (LVDT)

An LVDT (linear variable differential transformer) is an electromechanical sensor used for linear displacement measurements. One can use LVDT in

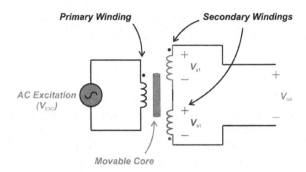

FIGURE 7.2   Schematic diagram of a typical LVDT sensor [10].

applications where displacements to be measured are evolving from a fraction of millimeters to a few centimeters. LVDT sensors are frequently used in structural monitoring applications such as recording displacement on structural members due to living loads and temperature variations. LVDT consists of a hollow metallic casing in which a shaft, called the core, moves freely back and forth along the axis of measurement. The core is made of a magnetically conductive material, and a coil assembly surrounds the metallic shaft. As shown schematically in Figure 7.2, the coil assembly consists of three transformer windings. A central primary winding is flanked by two secondary windings, one on either side. The outputs of the secondary windings are wired together to form a series of opposing circuits. When an AC excitation is applied to the primary winding, it generates an inductance current in the secondary windings, due to the mediation of the magnetically conductive core. When the core is equidistant between both secondary windings, no voltage appears at the secondary outputs. However, when the core moves, a differential voltage is induced at the secondary output. The magnitude of the output voltage changes linearly with the magnitude of the core's excursion from the center.

7.5.3.8   Accelerometer

An accelerometer is an electromechanical device used for measuring acceleration forces through single or multi-axis directions. Such forces can be static, like the continuous force of gravity on structural components, or dynamic to sense motions or vibrations like when a truck crosses a bridge. The application of accelerometers extends to multiple disciplines, from smartphones to rotating machinery and civil infrastructure. In the context of structural monitoring, accelerometers can be used for real-time monitoring of the variations of structural dynamic characteristics due to damage or changes in structural performance. The multi-axis models of accelerometers are mainly used to detect the magnitude and direction of the proper acceleration. Accelerometers have also wide use in constructions where there is a need to control the dynamic behavior of the structure, either short or long-term. Under structural applications, measuring and recording the dynamic behavior of structures is critical for evaluating their safety and viability. Dynamic loads can originate from a variety of sources such as impact loads (e.g., falling debris), construction work (e.g., driving piles, demolition), traffic loads,

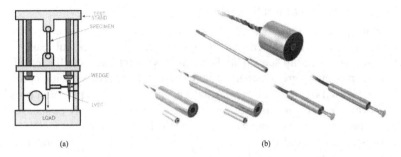

(a)   (b)

FIGURE 7.3 Schematic diagram. (a) Piezoelectric accelerometer. (b) A typical LVDT sensor [11].

earthquakes, and so on. Accelerometers used for civil engineering applications are either piezoelectric accelerometers or spring-mass accelerometers.

7.5.3.9 Piezoelectric accelerometer

A piezoelectric accelerometer is an accelerometer that employs the piezoelectric effect of certain materials (e.g., quartz) to generate electric charge in response to applied mechanical stress. This type of vibration transducer offers a very wide frequency and dynamic range. Piezoelectric devices are widely used in different industries, environments, and applications, allowing to measurement of dynamic changes in mechanical variables including acceleration, shock, and vibration. Piezoelectric accelerometers are light and small and operate over wide acceleration and frequency ranges. On the other hand, spring-mass accelerometers are relatively bulky and operate over a limited range of accelerations and frequencies [10]. However, they are very sensitive to small accelerations and provide better resolution than piezoelectric accelerometers. The piezoelectric accelerometer is made of a piezoelectric crystal element and an attached mass that is coupled to a supporting base. When the supporting base undergoes movement, the mass exerts an inertia force on the piezoelectric crystal element. The exerted force produces a proportional electric charge on the crystal (Figure 7.3). Since the force is equal to mass times acceleration, the charge is proportional to acceleration. The spring-mass accelerometer is essentially a damped oscillator.

7.5.3.10 Load Cells

A load cell is a type of transducer used for converting a mechanical force such as tension, compression, pressure, or torque into a measurable electrical output. This output changes proportionally to the force applied to the load cell. Load cells have been employed for a variety of applications that demand accuracy and precision. These sensors are employed in many historic buildings, where various building materials such as stone and brick have been used [11].

7.5.3.11 Temperature Sensors:

Temperature can be measured by a diverse array of sensors. Three types commonly used for civil engineering applications are: resistive; vibrating wire; and fiber optic temperature sensors [11].

7.5.3.12   Resistive Temperature Sensors

Resistive temperature sensors are based on the fact that the electrical resistance of a material changes as its temperature changes. There are two types of resistive temperature sensors: metallic sensors and thermistors. A typical metallic sensor comprises a fine platinum wire wrapped around a mandrel and covered with a protective coating or encased in a protective housing. The variation of the platinum resistance with temperature is linear. This variation can easily and accurately be measured by installing the sensor in one arm of the Wheatstone bridge circuit.

Thermistors are based on resistance change in a ceramic semiconductor. The resistance temperature relationship of a thermistor is negative and highly nonlinear. However, this difficulty is resolved by using thermistors in matched pairs in such a way that the nonlinearities of the two semiconductors offset each other. The operation range of thermistors is smaller than that of metallic temperature sensors, but thermistors usually provide higher accuracy. Resistive sensors all have a very important limitation. The current for the operation of these sensors, even though very small, creates a certain amount of heat, leading to an erroneous temperature reading.

7.5.3.13   Vibrating Wire Temperature Sensors

Vibrating wire temperature sensors operate similarly to VW strain gauges. A change in temperature causes a change in the frequency signal output from the VW temperature sensor. The readout device processes the signal and converts it to a voltage proportional to the temperature, or displays a reading in temperature units. The VW temperature sensor is encased in a cylinder to prevent physical contact between the sensor and the material. Therefore, no special precaution is needed for the effect of the strains on the reading of the sensor.

7.5.3.14   Fiber Optic Sensing Technology

In recent years, the demand for monitoring large civil infrastructures and the harsh conditions of monitoring environments prompted the development of new sensor technologies, and sensors have been under great development. In civil engineering, these sensors can be used to measure different parameters. Examples include strains, structural displacements, vibration frequencies, acceleration, pressure, temperature, and humidity. The structure monitoring can be either local or global. The focus of the local approach is on the material behavior while the global approach is attributed to monitoring the whole structural performance. Fiber optic sensors have been tested for different applications such as strain monitoring of concrete components in a bridge. Figure 7.4 shows Fiber Optic Sensing Technology.

Advantages of fiber optic sensors (FOSs): Fiber optic sensors are used primarily to measure variations in strain and/or temperature. Compared to conventional strain gages, FOSs offer the following advantages: (1) Stability: light signals can be transmitted along very long lengths with a very low signal transmission loss, allowing remote monitoring. FOSs are free from corrosion, have long-term stability, and allow continuous monitoring; (2) Non-conductive: FOSs are free from electromagnetic and radio frequency interferences, avoiding undesirable noise; (3) Convenience: FOSs and cabling are very small and light, making it possible to permanently incorporate them into the structures. Long-gage sensors are available for distributed sensing and the sensors can be virtually applied to any structural shape.

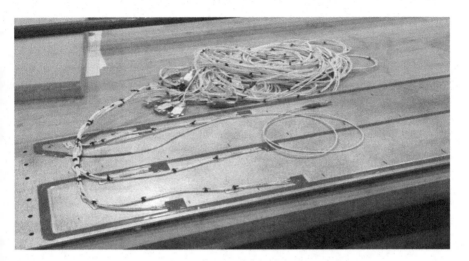

FIGURE 7.4   Fiber optic sensors [11].

7.5.3.15   Types of Fiber Optic Sensors

Since FOSs work by measuring changes in the physical properties of the guided light, a number of different fiber optic sensors have been developed in recent years based on different modulation of properties of the light sensing techniques. Intensity, interferometric, and spectrometric sensors are categorized by their transduction mechanisms while localized, multiplexed, and distributed sensors are by their applications.

a. Intensity sensors: Sensors based on intensity modulation pertain to light intensity losses that are associated with straining of optical fibers along any portion of their length (see Figure 7.5). The advantages of intensity or amplitude-type sensors are the simplicity of construction and compatibility with multi-mode fiber technology. The drawbacks are measurements are only relative and variations in the intensity of the light source may lead to false readings unless a referencing system is used.

b. Spectrometric sensors: Spectrometric sensors are based on relating the changes in the wavelength of light to the measurement of interest, that is, strain. An example of such a sensor for measuring strains is the Bragg grating sensor [13] (see Figure 7.6). In a fiber Bragg grating (FBG) sensor, an optical grating (essentially a series of tiny reflectors) is placed on the fiber and the grating spacing is proportional to the wavelength of light reflected when a light pulse is sent down the fiber. When strain is induced at the location of the grating, it causes this grating spacing to change, and this causes a shift in the wavelength of the reflected light. Through the use of a specialized optical technique, along with analysis and calibration of the FBG, the data obtained from the grating spacing can be converted to a measured strain value. FBG sensors are commercially available and are intended to measure local "point" strains only. They can be used for both static and dynamic monitoring and can be serially multiplexed. These sensors have

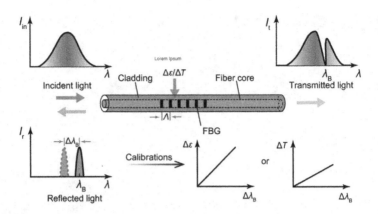

FIGURE 7.5 Optical fiber intensity sensors [12].

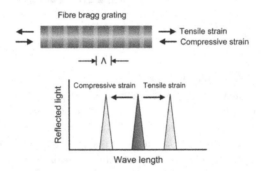

FIGURE 7.6 Strain-induced shift in wavelength for a fiber Bragg grating [12].

successfully been embedded within construction materials, and they are bondable and weldable. However, it should be noted that FBG sensors are sensitive to temperature and require thermal compensation during data collection. FBG sensors are sensitive to extremely small strains. Fiber optic sensors based on FBG technology are suitable for direct strain and temperature sensing (as well as indirect measurement) and have a number of advantages compared to conventional strain gauges.

c. Interferometric sensors: Interferometric sensors can be configured in a number of different ways for sensing purposes (see Figure 7.7). Interferometric sensors require the interference of light from two identical single-mode fibers, one of which is used as a reference arm, and the other is the actual sensor. An exception to a two-arm interferometric sensor is a single–fiber Fabry–Perot type sensor [14]. In a Fabry–Perot type sensor, the fiber is manipulated with two parallel reflectors (mirrors) perpendicular to the axis of the fiber. The interference of the reflected signals between the two mirrors creates the interference pattern. The interference pattern generated at the output end of the phase sensors is directly related to the intensity of the applied strain field between the two reflectors.

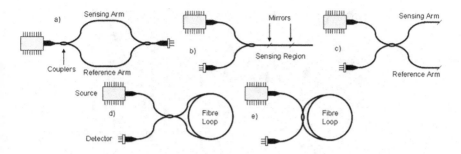

FIGURE 7.7 Fiber optical interferometric sensors. (a) Mach Zender. (b) Michelson. (c) Fabry Perot [12].

d. Brillouin scattering sensor: Brillouin scattering-based sensing systems are still in the developmental stage. When light travels through a transparent media, most of it travels straight forward, as a small fraction is backscattered. Different components of the backscattered light can be identified. A Brillouin scattering sensor exploits the sensitivity of the Brillouin frequency shift for temperature and strain sensing applications. The technique uses standard low-loss single-mode optical fiber offering the longest distance range with unrivalled performances and compatibility with standard telecommunication components. In this technology, Brillouin scattering is usually optically stimulated leading to the greatest intensity of the scattering mechanism and consequently an improved signal-to-noise ratio. Brillouin's frequency-based technique is opposed to intensity-based techniques and is inherently more accurate and more stable in the long term since intensity-based techniques suffer from a high sensitivity to drifts. Brillouin scattering sensor measures static strain profiles using a single optical fiber. This means that these sensors can be used to measure the distribution of strains along their length, a somewhat unique capability. The gauge length of these sensors can vary from 15 cm to more than 1,000 m. The Brillouin scattering wavelength shift is dependent upon temperature and strain. To function as a strain sensor, it must be configured to discriminate between strain and temperature in a manner similar to the approach taken for Bragg grating sensors. Use of this type of sensor requires extensive analysis of optical signals and data, and at present Brillouin Scattering FOS systems are very expensive. Of the sensor types discussed above, intensity-type sensors are simple to construct but their sensitivity is rather low. Interferometric sensors offer the highest sensitivity but the required components are quite complicated. The most commercially available FOSs are FBG sensors and Fabry–Perot sensors.

7.5.3.16 Strain Gauges in Structural Health Monitoring

The most common type of load cells used in structural monitoring are strain gauges as shown in Figure 7.8. A strain gauge is a device used to measure strain due to applied force on an object. The most common type of strain gauge consists of an insulating flexible backing that supports a metallic foil pattern. The gauge is attached

FIGURE 7.8   Strain gauges monitoring sensors [11].

to the object by a suitable adhesive material. When subjected to force, the foil is deformed, causing its electrical resistance to change which can then be measured. These sensors are most often used to monitor strain in steel and reinforced concrete structures [15]. A strain gauge rosette is one type of strain gauge composed of two or more strain gauges that are positioned closely to measure strains along different directions of the component under evaluation. The use of multiple strain gauges enables a more precise evaluation of strain on the surface being measured.

7.5.3.17   Inclinometer (Slope Indicator)

Inclinometers (also known as slope indicators) are precision instruments used to monitor subsurface movements and deformations. Inclinometers are designed to measure horizontal subsurface deformation in a borehole when slope stability is a concern for natural slopes, constructed cut/fill slopes, and deep-fill projects. An inclinometer system contains two main components: an inclinometer casing and an inclinometer measurement probe. The inclinometer measurement probe is lowered and raised through a specialty inclinometer casing with precision machined grooves which control the orientation of the sensor and provide a uniform surface for measurements. Inclinometers are usually installed in a borehole; however, they can also be buried in a trench (horizontal inclinometers), cast into concrete, or attached to a structure. Inclinometers are typically used for (1) Determining whether subsurface movements are constant, accelerating, or responding to remedial measures. (2) Checking that subsurface deformations are within design limits, that struts and anchors are performing as expected, and that neighboring buildings are not affected by ground movements. (3) Monitoring settlement profiles of embankments, foundations, and other structural components (horizontal inclinometer) [11].

7.5.3.18   Tiltmeter

A tiltmeter is a sensitive inclinometer used for monitoring very small changes in the inclination of a structure. The measured data provide an accurate history of the movement of a structure, which can be used for early warning of potential structural damage. The sensors are able to measure angles of slope (or tilt), elevation, or depression of an object with respect to gravity's direction. These sensors are suited for monitoring rotation of structures such as concrete dams or retaining walls.

7.5.3.19 Acoustic Emission Sensor

Acoustic Emission (AE) Sensors measure high-frequency energy signals that are generated from local sources of stress waves. Discontinuities and defects in a material generate stress waves. AE sensors are able to pick up the stress waves propagated to the material's surface. By converting these waves into electrical signals, AE sensors are ideal devices to effectively assess the current state of materials under stress. These sensors are mainly used to detect the onset or growth of existing cracks in structural components [11].

7.5.3.20 Temperature Sensors

Civil engineering structures are subject to environmental changes and therefore it is necessary to measure the temperature that affects the physical properties of structures to some extent [15]. Thermocouples are one of the most widely used temperature sensors to measure the variations of temperature in certain points of the structure. Most of the large concrete structures use temperature sensors, while casting and during construction, in order to have full control over temperature changes under curing.

7.6 FUSION AND INTEGRATION: THE DIFFERENCE

Let's start with the data fusion definition. Both data fusion and data integration are designed to integrate and organize data that comes from multiple sources. The goal is to present a unified view of data for consumption by various applications, making it easier for analytics to derive actionable insights. However, there are major differences between data fusion and data integration. The main one is that information fusion focuses on processing real-time streaming data and enriching this stream with semantic context from other Big Data sources. Other differences include:

- Data Reduction
- The first and foremost goal of information fusion is to enable data abstraction. So, data integration focuses on combining data to create consumable data. Information Fusion frequently involves "fusing" data at different abstraction levels and various levels of uncertainty to support a more narrow set of application workloads.

7.6.1 Handling Streaming/Real-Time Data

Data integration is often combined with data-at-rest or batch-oriented data. Information Fusion integrates, transforms, and organizes all manner of data (structured, semi-structured, and unstructured) and uses multiple techniques to reduce the amount of stateless data and only retain stateful and valuable information. Human Interfaces Information Fusion incorporates contributions to the data and reduces uncertainty. If organizations can add and save interfaces that can only be derived with human analysis and support, they will be able to maximize their analytics results.

7.6.2 PERCEPTION SENSORS

One of the most important tasks of autonomous systems is to acquire knowledge about their environment. This is done by taking measurements using various sensors and then developing inferences from those measurements. An autonomous vehicle may experience unforeseen events on the road which it needs to register and act accordingly. Although our focus will be on multi-sensor data fusion, single-sensor perception has been widely studied in the literature. Perception sensors can be classified into two types: active and passive sensors. Active sensors emit their own energy into the environment and then measure the environmental reaction. For object detection, a few different types of active sensors can be used:

7.7 TYPES OF SENSORS FOR TARGET DETECTION AND TRACKING

The ultimate goal when a robot is built is to be optimized and to be compliant with all specifications. To meet the requirements sometimes you can spend many hours just to sort and identify the sensors that would be the best for an application like detecting and tracking an object. In this article, we explore all sensor types that can be used for target detection and tracking as well as features and the types of applications where they can be used. Selecting the right sensor is not a strict process. This is about eliminating all the wrong choices based on a series of questions aiming to eliminate first the technology underlying the sensor and then the product that doesn't fit the robot's requirements. When we use the word target, we refer to the same time as a small ball, an object like a chair, or even a human that stays in front of the robot. To select the best sensor from a variety of products and manufacturers is hard work especially when you're a beginner and try to build a simple robot. In a few words, the sensor has to be selected in concordance with your target's size, shape, and range. All of these three features have to be on the same line with the specification of the robot. But even so, it is hard to define the best sensor since the performance and precision of this depend on many factors.

A sensor is a sophisticated device that measures a physical quantity like speed or pressure and converts it into a signal that can be measured electrically. Sensors are based on several working principles and types of measurements. In our case, almost all types of sensors emit signals and measure the reflection to make measurements. There are many sensors that can be used for a simple application like line following (IR LED and a Photodiode, LED and LDR, etc.), but this is a simple case when a simple sensor can be selected. A complicated case is when you have to track an object and the budget is limited to purchase a mini computer like Raspberry Pi. In this case, it can be used an ultrasonic sensor to scan from side to side till the sensor detects a drop in distance (at this stage it detects the edge of the object, and from now on the sensor sees only the background). The scanning process continues back to the point where the object is with left and right scanning. This is one of the cases when expensive products can be replaced with cheap sensors. In the following, you can explore the features of sensors that can be used for detecting and tracking a target.

7.8 CRITERIA TO CHOOSE THE RIGHT SENSOR

With so many models available on the market from various vendors would be very useful to know from where you can start. A good starting point is to know certain features and some of them are given below:

- Type of sensor: the presence of an object can be detected with proximity sensors, and there are several kinds of sensor technologies including here ultrasonic sensors, capacitive, photoelectric, inductive, or magnetic. Tracking objects can work using proximity sensors (e.g., ultrasonic sensors), or for advanced applications generally it uses image sensors (e.g., webcams) and vision software like OpenCV;
- Accuracy: accuracy is very important in detecting and tracking objects, and it is useful to choose sensors with accuracy values between desired measurement margins;
- Resolution: a high resolution can detect the smallest changes in the position of the target;
- Range: involves choosing the sensors based on measurement limits and compared with the desired detection range of the robot;
- Control interface: to interface the sensor you have to know the types of the sensors. A wide range of sensors are three-wire DC types, but there are many more types including two-wire DC or two-wire AC/DC;
- Environmental condition: any sensor has its operational limits usually these are the temperature and humidity;
- Calibration: calibrating the sensors is an essential step to ensure efficiency and more accurate measurement;
- Cost: depending on the budget allocated to a project, you can select the sensor or the sensors that can be used to build the robot;

7.9 SENSOR CLASSIFICATION BASED ON PROPERTY

- Proximity sensors: several sensor technologies are used to build proximity sensors: ultrasonic sensors, capacitive, photoelectric, inductive, or magnetic;
- Motion detectors: these sensors are based on infrared light, ultrasound, or microwave/radar technology;
- Image sensors: these are digital cameras, camera modules, and other imaging devices based on CCD or CMOS technology;

Starting from the above information and combining the below features for each type of sensor separately, you are now able to find the appropriate sensor and start.

7.10 BUILDING ANY ROBOTIC APPLICATION

Each sensor type has its characteristics that make this little device better for a certain task or replaceable for other tasks. For example, an ultrasonic sensor works fine for solid objects and becomes lazy for soft or fuzzy objects. Also, some sensors are unable to make the difference between a static object and a human. All of these characteristics have to be clear before choosing the right sensor/sensors for your robot.

7.10.1 LIGHT SENSOR

A light sensor can be included in the proximity sensor category, and it is a simple sensor that changes the voltage of Photoresistor or Photovoltaic cells in concordance with the amount of light detected. A light sensor is used in very popular applications for autonomous robots that track a line-marked path.

7.10.2 COLOR SENSOR

Different colors are reflected with different intensities, for example, the orange color reflects red light in an amount greater than the green color, and this is the color sensor. This simple sensor is in the same range as a light sensor, but with a few extra features that can be useful for applications where the robot has to detect the presence of an object with a certain color or to detect the types of objects or surfaces.

7.10.3 TOUCH SENSOR

The touch sensor can be included in the proximity sensors category and is designed to sense objects at a small distance with or without direct contact. This sensor is designed to detect the changes in the capacitance between the on-board electrodes and the object making contact.

7.10.4 ULTRASONIC SENSOR

These sensors are designed to generate high-frequency sound waves and receive the echo reflected by the target. These sensors are used in a wide range of applications and are very useful when it is not important for the detection of colors, surface texture, or transparency.

Advantages of ultrasonic sensors:

- the output value is linear with the distance between the sensor and the target;
- sensor response is not dependent on the colors, transparency of objects, optical reflection properties, or the surface texture of the object;
- these sensors are designed for contact-free detection;
- sensors with digital (ON/OFF) outputs have excellent repeat sensing accuracy;
- accurate detection even of small objects;
- ultrasonic sensors can work in critical conditions such as dirt and dust;
- they are available in cuboid or cylinder forms, which is better for a freedom design;

Disadvantages of ultrasonic sensors:

- ultrasonic sensors must view a high-density surface for good results. A soft surface like foam and cloth has a low density and absorbs the sound waves emitted by the sensor;
- could have false responses for some loud noises such as air hoses;

- the ultrasonic sensors have a response time of a fraction less than other types of sensors;
- an ultrasonic sensor has a minimum sensing distance, which should be taken into consideration when you choose the sensor;
- some changes in the environment can affect the response of the sensor (temperature, humidity, pressure, etc.);

7.10.5 INFRARED SENSOR

An infrared sensor measures the IR light that is transmitted in the environment to find objects by an IR LED. This type of sensor is very popular in navigation for object avoidance, distance measurement, or line-following applications. This sensor is very sensitive to IR lights and sunlight, and this is the main reason that an IR sensor is used with great precision in spaces with low light.

Advantages of ultrasonic sensors:

- infrared sensors can detect infrared light over a large area;
- they can operate in real time;
- the IR sensor uses non-visible light for detection;
- they are cheap sensors;

Disadvantages of ultrasonic sensors:

- this sensor is very sensitive to IR lights and sunlight;
- it has a weakness to darker colors such as black;

7.10.6 SONAR SENSOR

The sonar sensor can be used primarily in navigation for object detection, even for small objects, and generally is used in projects with a big budget because this type of sensor is very expensive. This sensor has high performance on the ground and in water where it can be used for submersed robotics projects.

7.10.7 LASER SENSOR

A laser light is very useful for tracking and detecting a target located at a long distance. The distance between the sensor and the target is measured by calculating the speed of light and the time since light is emitted and until it is returned to the receiver.

A laser sensor is very precise in measurement and at the same time is very expensive.

7.10.8 IMAGE SENSOR

The most popular combination for detecting and tracking an object or detecting a human face is a webcam and the OpenCV vision software. This combination may be the best in detection and tracking applications, but it is necessary to have advanced programming skills and a mini-computer like a Raspberry Pi. Using an image sensor can be built a wide range of applications, and some of these are listed below:

- face detection and tracking;
- tracking and detecting objects in colors;
- detect specific shapes in images;
- detect corners of triangles from an image;
- detect the position of an object on 2D surfaces;
- it can acquire and prioritize targets;

7.11 CRITERIA TO CHOOSE THE RIGHT SENSOR

Regardless of the technique used to monitor a structure, whether it is vibration-based, strain-based, ultrasonic-based, etc. [6], or combined, there are some common criteria that must be followed when designing a sensor network.

- With so many choices available in the market from various vendors, it would be really helpful to know where to start. A good starting point is to know certain features; some of which are listed below:
- Type of sensor: The presence of an object can be detected with proximity sensors and other sensor technologies like ultrasonic sensors, capacitive, photoelectric, inductive, or magnetic; or for advanced applications, generally image sensors and vision software like OpenCV are used.
- Accuracy: Accuracy is a very important element, and it is useful to choose sensors with accuracy values between desired measurement margins.
- Resolution: Depending on the type of objects to be tracked, we can choose sensors based on resolution. A high resolution can detect even the smallest changes in the position of the target.
- Range: This involves choosing the sensors based on measurement limits.
- Control interface: To interface the sensor you have to know the types of the sensors. A wide range of sensors are three-wire DC types, but there are many more types, including two-wire DC or two-wire AC/DC;
- Environmental conditions: These include sensor operational limits like temperature and humidity.
- Calibration: Calibrating the sensors is an essential step to ensure accurate measurement and efficiency.
- Cost: Depends on a number of factors the sensor is supporting.
- Sensor classification based on property
- Proximity sensors: Several sensor technologies are used to build proximity sensors: ultrasonic sensors, capacitive, photoelectric, inductive, or magnetic;
- Light sensor: It is a simple sensor that changes the voltage of Photovoltaic cells in concordance with the amount of light detected. A light sensor is used in very popular applications that track a line-marked path.
- Color sensor: Different colors are reflected with different intensity. This simple sensor is in the same range as a light sensor, but with a few extra

features that can be useful for applications where the device has to detect the presence of an object with a certain color.

- Touch sensor: The touch sensor can be included in the proximity sensors category and is designed to sense objects at a small distance with or without direct contact. This sensor is designed to detect the changes in the capacitance between the onboard electrodes and the object making the contact.
- Ultrasonic sensor: These sensors are designed to generate high-frequency sound waves and receive the echo reflected by the target. These sensors are used in a wide range of applications and are very useful when it is not important to detect colors, surface texture, or transparency.
- Motion detectors: These sensors are based on infrared light, ultrasound, or microwave/radar technology.
- Infrared sensor: An infrared sensor measures the IR light that is transmitted in the environment, to find objects by an IR LED. This type of sensor is very popular in navigation for object avoidance, distance measurement, or line-following applications.
- Sonar sensor: The sonar sensor can be used primarily in navigation for object detection, even for small objects, and generally is used in projects with a big budget, because this type of sensor is very expensive.
- Laser sensor: A laser light is very useful for tracking and detecting a target located at a long distance. The distance between the sensor and the target is measured by calculating the speed of light and the time since light is emitted and until it is returned to the receiver.

A laser sensor is very precise in measurement and at the same time is very expensive.

- Image sensors – these are digital cameras, camera modules, and other imaging devices based on CCD or CMOS technology.
- Each IoT sensor type has its characteristics that can help make the device better for a certain task or replaceable for other tasks. For example, an ultrasonic sensor works fine for solid objects but becomes lazy for soft or fuzzy objects. Also, some sensors are unable to Figure 7.9 the difference between a static object and a human being. All of these characteristics have to be clear before you choose the right sensor/sensors for your object detection solution. So, with the help of

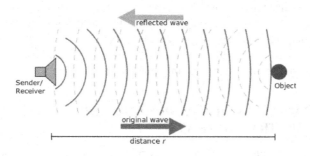

FIGURE 7.9   Principle of an active sonar. Photo source wikipedia.org.

comparative information on each type of sensor separately, you will now be able to find the appropriate sensor for any object-tracking solution.

- Background and Development
- Multi-sensor integration is a system-controlled process based on integrating information from different sensors when a multi-sensor system completes a task. It emphasizes the conversion of different data and the overall flow structure for the system. Multi-sensor fusion is a specific stage in the process of multi-sensor integration, in which sensor information is merged into unified comprehensive information. The specific methods and procedures of data conversion and merger are emphasized.
- Multi-sensor integration and fusion technology is actually a comprehensive technology of multi-source information. In multi-sensor integration and fusion, the most consistent estimate of a measured object and its nature is obtained through analysis and comprehension of data information from various sensors.
- Multi-sensor fusion research started as early as 1973, when research institutions in the United States, funded by the Department of Defense, started to study sonar signal understanding systems. From then on, multi-sensor fusion technology developed rapidly. Apart from being used in the C 3I (Command / Control / Communication Intelligence) system, many kinds of sensors are now widely used to gather information in industrial control, robotics, air traffic control, marine surveillance, and management Multi-sensor fusion research has become an issue of concern in the military, production and high-technology development areas.
- In 1988, the U.S. Department of Defense listed the C3I multi-sensor fusion technology as one of 20 key technologies of research and development in the 1990s. The C3I expert group led by the U.S. Secretary of Defense set up a specialized data fusion group and organized a number of thematic plans to study multi-sensor fusion technology, and since 1992 some 100 million U.S. dollars annually have been invested in multi-sensor fusion technology. Related journals are also published and specialized conferences are held internationally so that there is a well-researched environment for multi-sensor fusion technology.
- Multi-sensor fusion theory is divided into numerical approach research of similar information fusion, and symbol approach research of information integration of different types. Numerical approach research of similar information fusion, especially distribution of all kinds of optimal, sub-optimal, or partly dispersed algorithms, is more dominant, while symbol approach research of information integration of different types is more difficult in theoretical research mainly oriented to exploratory research. There are many application systems with the feature of multi-sensor information integration, such as the early warning integration.

7.12   INSTALLATION OF SENSORS AND DATA ACQUISITION SYSTEMS

Installation of the sensors and devices is performed according to the installation plan. The size of the installation team depends on the size of the project.

The work to be done before the actual installation is the following:

- The sensors and devices to be installed are delivered, checked, and quality controlled.
- The persons installing the sensors are well-informed about the installation procedure and need even to have a good understanding of the sensor devices. When installing new complicated sensors like for example fiber optic sensors it is necessary to give the new personnel a short course about the sensors and installation procedure.
- The necessary equipment is procured and if necessary tested.
- Data acquisition systems are calibrated and the software is programmed and tested in the office.
- A lot of practical details often delay the project and therefore it is good to check the following things: access and keys to the building site, safety regulations, other activities that might collide with the installation at the building site, access to electricity and facilities for the personnel, etc.
- Passage for the cables etc. is done beforehand if possible. This is especially important in new concrete constructions where the plastic pipes or such need to be concreted beforehand.

The work to be done in the actual installation is the following:

- The sensors and devices are installed according to the installation plan and drawings
- If any changes are made, they are noted so that the drawings can be revised
- The installation procedure is described in detail in a diary and documented with photographs
- The sensors are tested, measured, and calibrated if possible or necessary
- If there is a risk for damages, the sensors are protected or marked
- The sensors are connected to the data logger for temporary or permanent measurement or if they are not to be used directly, they are protected or set in a safe place
- If temporary measurements are performed during some stage of the construction the measurement equipment is protected against damage and marked clearly
- In the case of a new concrete construction embedded sensors it is good to be present when concreting in order to supervise the survival of the sensors

The work to be done after the installation of the sensors and cables is the following:

- The sensors and devices are connected to connection boxes, main units, etc. according to the installation plan and drawings
- The communication system is established
- The cables are fixed temporarily or permanently on the cable rack
- If any changes are made, they are noted so that the drawings can be revised

- The installation procedure is described in detail in a diary and documented with photographs
- The system is tested
- The system and monitoring are verified by other systems, models, or calculations

Every project has unique requirements and therefore it is not easy to describe a procedure that covers all details. Good and very detailed planning saves time and thereby the costs for the installation and ensures qualified installation.

7.13   DESIGN REQUIREMENTS FOR SENSOR NETWORKS

Regardless of the technique used to monitor a structure, whether it is vibration-based, strain-based, ultrasonic-based, etc. [16], or combined, there are some common criteria that must be followed when designing a sensor network. In this section, we will identify the basic requirements and challenges, and further, we will present a short case study. The selection of the type and the size of the sensor elements/nodes in the network is a major component that needs to be decided upon before considering other design requirements within a sensor network. The type of sensors highly depends on the application and the structure to be monitored. For instance, when using an accelerometer to measure the level of vibration, the interest will be more on understanding the global behavior of the structure [17], that is, the shift in the modal frequencies that may be used to identify the presence of damage. Vibration-based methods have been implemented already on many structures, mainly civil structures, albeit they lack a qualitative assessment of the health state of the structure [18]. On the other hand, when strain sensors (using strain gauges or fiber optic sensors) or ultrasonic transducers are considered, the focus will be more on the local assessment of any damage that may be present within the structure [19]. These techniques can detect small surface and embedded defects such as corrosion, fatigue cracking, and impact damage [20]. Energy supply for the sensors in the network is mandatory for its reliable and efficient operation. With the presence of an energy source, powering the sensor will not be an issue despite the level of energy needed and the operation time. However, in the absence of an energy source, in particular when monitoring structures in rural areas, an alternative energy source must exist. Internal batteries may be an option, keeping in mind the limited lifetime that the batteries have before needing to be replaced. The energy consumption of the node, when an internal battery is used, brings several challenges such as the duration and frequency of data collection. A microcontroller can regulate the sleep and wake-up time for the sensor during or when an incident happens such as traffic or crowd loading on a bridge, wind load on a high-rise building, or a bird strike on an aircraft nose. An alternative source for powering sensors, that has attracted the attention of many researchers working in the area of SHM, relies on the harvesting of various forms of green energy such as light using photovoltaic cells and kinetic energy using piezoelectric materials or electromagnetic devices [21]. When it comes to data communication, sensors can be wired or wireless depending on the need. Normally wired sensors acquire data using an acquisition system and the data is transmitted to an on-site PC through the LAN. Later the data can be transmitted to a central server [22]. The onsite computers may use standard TCP/IP communication protocols and therefore can communicate across a wide area network such as the Internet using standard equipment including routers and secure VPNs. Onsite can be an optional network services computer to provide NTP services (if very accurate time synchronization between sensing nodes is required) and onsite data caching. Wired sensors may not be ideal when dealing with large structures such as pipelines due to the complexity of the wiring system. The development in wireless sensing technologies led to a major advancement in the field of SHM in particular when instrumenting large and complex structures, due to their advantages related to the ease of deployment and the ability to do local processing. Many hardware has been developed to satisfy the needs of SHM systems (such as NI, HBM, LORD Sensing MicroStrain, etc.). Besides, many communication protocols have been developed and proven to be reliable for SHM applications based on the IEEE 802.15.4 communication standard [23]. This will be elaborated on more in the later sections. Data transmission relies highly on the type of sensors used. Continuous data transmission is plausible when wired sensors are used, yet the challenge becomes in the data management and storage, as well as the ability to extract the indicative features that can be used for structural or operational assessments. On the other hand, wireless sensors are often powered using embedded small batteries, and therefore, energy consumption should be minimized. This is usually achieved by reducing radio communication via controlling the duty cycle, as well as the in-network processing. Duty cycles focus on the sleep and wake-up time of the sensors, while in-network processing focuses on the amount of data to be transmitted that may be achieved through data compression [24]. Data loss during transmission, resulting from data compression, is a common issue and must be well thought out [25]. Various types of lossless data compression algorithms are available including Huffman's coding, Run Length encoding, Dictionary coders (LZW), etc. [26]. Data transmitted is classified as confidential, and therefore, security issues with wireless sensor network (WSN)-based SHM systems must be addressed. Many research scientists and engineers have tackled the issues of eavesdropping, traffic analysis, disruption of the sensor application, or hijacking [27]. ZigBee has proven to be an effective, feasible, and reliable wireless sensor technology for application in SHM [28]. In terms of security, ZigBee incorporates all the security mechanisms proposed by the IEEE 802.15.4 (such as message encryption) [29]. Moreover, given that ZigBee has been developed to support lower data rates and low power transmission, the increase in the number of nodes can be successfully implemented and devices run for years on inexpensive batteries. Once the data has been transmitted to the servers, it must be stored efficiently for ease of access when queries are executed. Therefore, an appropriate database tool must be selected. Relational database management systems and structures query language (SQL) have been widely used and implemented in many SHM systems due to their reliability and extensive user base. RDBMS may suffer in terms of writing and reading, and scalability, hence NoSQL (not only SQL) have been proposed. NoSQL is known for its advanced performance and ability to support more data schema [30]. Finally come the data analytics and interpretation of data, various techniques, and methods based on signal processing and data-driven models have been developed for this purpose. This component of the SHM system highly depends on the nature of the structure, the type of data collected, and most importantly the sensitivity of the measurements to a given fault or malfunction in the operation of the system monitored. The designer of a sensor network needs to pay attention to the main requirements of a robust network which mainly involves the sensor selection, placement of sensors, methods to power sensors, data transmission, storage, and data analytics. The integration of existing technology is a major challenge to achieve an efficient sensing system, hence efforts are required to develop sensor nodes that can provide different types of measurements, also nodes that are self-powered and can withstand severe weather conditions.

REFERENCES

1. Bojan, T. M., Kumar, U. R., & Bojan, V. M. (2014). An internet of things-based intelligent transportation system. In *2014 IEEE International Conference on Vehicular Electronics and Safety* (pp. 174–179). Hyderabad, India. doi: 10.1109/ICVES.2014.7063743.
2. Brincat, A. A., Pacifici, F., Martinaglia, S., & Mazzola, F. (2019). The internet of things for intelligent transportation systems in real smart cities scenarios. In *2019 IEEE 5th World Forum on Internet of Things* (WF-IoT; pp. 128–132). Limerick, Ireland. doi: 10.1109/WF-IoT.2019.8767247.
3. Zantalis, F., Koulouras, G., Karabetsos, S., & Kandris, D. (2019). A review of machine learning and IoT in smart transportation. *Future Internet*, 11, 94.
4. Singh, D., Tripathi, G., & Jara, AJ. (2014). A survey of Internet-of-Things: future vision, architecture, challenges and services. In *2014 IEEE World Forum on Internet of Things (WF-IoT)* (pp. 287–292). March 6–8, Seoul, South Korea, IEEE.
5. Luan, T. H., Gao, L., Li, Z., Xiang, Y., & Sun, L. (2015) Fog Computing: Focusing on Mobile Users at the Edge. Available from: https://arxiv.org/abs/1502.01815 (accessed on 28 February 2022).
6. Dogra1, A. K., & Kaur, J. (2022). Moving towards smart transportation with machine learning and Internet of Things (IoT): a review. *Journal of Smart Environments and Green Computing*. Available from: https://api.semanticscholar.org/CorpusID:247213146.
7. Burgos, D. A. T., Vargas, R. C. G., Pedraza, C., Agis, D., & Pozo, F. (2020). Damage identification in structural health monitoring: a brief review from its implementation to the use of data-driven applications. *Sensors*, 20, 733.
8. Leon-Medina, J. X., Anaya, M., Pozo, F., & Tibaduiza, D. (2020). Nonlinear feature extraction through manifold learning in an electronic tongue classification task. *Sensors*, 20, 4834.
9. Mufti, A. (2001). Guidelines for Structural Health Monitoring. ISIS Canada Design Manual No. 2 September 2001. Available from: https://www.scribd.com/document/368478048/Guidelines-for-Structural-Health-Monitoring
10. Dong, Y. (2010). *Bridges Structural Health Monitoring and Deterioration Detection Synthesis of Knowledge and Technology*. Final Report. University of Alaska Fairbanks, Fairbanks, AK.
11. Sensors for Structural Health Monitoring. Available from: https://www.fprimec.com/sensors-for-structural-health-monitoring/.
12. Ansari, F. (1997). State-of-the-art in the applications of fiber optic sensors to cementitious composites. *Cement & Concrete Composites*, 19(1), 3–19.
13. Morey, W. W., Meltz, G., & Glenn, D. H. (1989). Fiber optic Bragg grating sensors. *Proceedings of SPIE Fiber Optic and Laser Sensors*, 1169, 98.
14. Claus, R. O., Gunther, M. F., Wang. A. B., Murphy, K. A., & Sun, D. (1993). Extrinsic fabry-perot sensor for structural evaluation. In F. Ansari (Ed.), *Applications of Fiber Optic Sensors in Engineering Mechanics, ASCE-EMD Spect.* (pp. 60–70). New York, ASCE.

15. Shao, X. P., Guo, B. L., & Ni, Y. Q. (2010). Instrumentation for reinforced concrete durability monitoring of Qingdao Bay Bridge. In Bridge Maintenance, Safety, Management and Life-Cycle Optimization – *Proceedings of the 5th International Conference on Bridge Maintenance, Safety and Management* (pp. 851–856). Philadelphia, PA.

16. Runcie, P., Mustapha, S., & Rakotoarivelo, T. (2014). Advances in structural health monitoring system architecture. In Furuta, H., Frangopol, D., & Akiyama, M. (eds.), *Proceedings of the Fourth International Symposium on Life-Cycle Civil Engineering, IALCCE* (pp. 1064–1071). 16–19 November 2014, Tokyo, Japan, CRC Press. https://doi.org/10.1201/b17618

17. Ostachowicz, W., Soman, R., & Malinowski, P. (2019). Optimization of sensor placement for structural health monitoring: a review. *Structural Health Monitoring*, 18, 963–988.

18. Ismail, Z., Mustapha, S., Fakih, M. A., & Tarhini, H. (2020). Sensor placement optimization on complex and large metallic and composite structures. *Structural Health Monitoring*, 19, 262–280.

19. Su, Z., Ye, L., & Lu, Y. (2006). Guided lamb waves for identification of damage in composite structures: a review. *Journal of Sound and Vibration*, 295, 753–780.

20. Mustapha, S., Braytee, A., & Ye, L. (2018). Multisource data fusion for classification of surface cracks in steel pipes. *Journal of Nondestructive Evaluation, Diagnostics and Prognostics of Engineering Systems*, 1, 021007–021007-11.

21. Anaissi, A., Khoa, N. L. D., Mustapha, S., Alamdari, M. M., Braytee, A., Wang, Y., & Chen, F. Adaptive one-class support vector machine for damage detection in structural health monitoring. In *Proceedings of the Pacific-Asia Conference on Knowledge Discovery and Data Mining* (pp. 42–57). 23–26 May 2017, Jeju, Korea.

22. Yang, J., He, J., Guan, X., Wang, D., Chen, H., Zhang, W., & Liu, Y. (2016). A probabilistic crack size quantification method using in-situ Lamb wave test and Bayesian updating. *Mechanical Systems and Signal Processing*, 78, 118–133.

23. Dong, W., Lin, Y., Zhongqing, S., Ye, L., Fucai, L., & Guang, M. (2009). Probabilistic damage identification based on correlation analysis using guided wave signals in aluminum plates. *Structural Health Monitoring*, 9, 133–144.

24. Musta, K., Alamdari, M. M., Runcie, P., Dackermann, U., Nguyen, V., Li, J., & Ye, L. (2015). Pattern recognition based on time series analysis using vibration data for structural health monitoring in civil structures. *Electronic Journal of Structural Engineering*, 14, 106–115.

25. Doebling, S. W., Farrar, C. R., & Prime, M. B. (1998). A summary review of vibration-based damage identification methods. *Shock and Vibration Digest*, 30, 91–105.

26. Yeager, M., Todd, M., Gregory, W., & Key, C. (2016). Assessment of embedded fiber Bragg gratings for structural health monitoring of composites. *Structural Health Monitoring*, 16, 262–275.

27. Mustapha, S., Ye, L., Dong, X., & Alamdari, M. M. (2016). Evaluation of barely visible indentation damage (BVID) in CF/EP sandwich composites using guided wave signals. *Mechanical Systems and Signal Processing*, 76–77, 497–517.

28. Tang, X., Wang, X., Cattley, R., Gu, F., & Ball, D. A. (2018). Energy harvesting technologies for achieving self-powered wireless sensor networks in machine condition monitoring: a review. *Sensors*, 18, 4113.

29. Chen, B., Wang, X., Sun, D., & Xie, X. (2014). Integrated system of structural health monitoring and intelligent management for a cable-stayed bridge. *The Scientific World Journal*, 2014, 12.

30. Gutierrez, J. A., Callaway, E. H., & Barrett, R. (2003). *IEEE 802.15.4 Low-Rate Wireless Personal Area Networks: Enabling Wireless Sensor Networks*. Middlesex County, NJ, IEEE Standards.

8 Selection of the Case Studies

8.1 BACKGROUND

Road transportation is the backbone of modern economies, albeit it yearly costs about 1.25 million deaths and trillions of dollars to the global economy, and damages public health and the environment [1,2]. Advances in Information and Communication Technologies (ICT) in areas such as hardware, software, and communications have created new opportunities for developing a sustainable, intelligent transportation system. The integration of ICT with the transportation infrastructure will enable a better, safer traveling experience and migration to Intelligent Transportation Systems (ITS) which focus on four fundamental principles: sustainability, integration, safety, and responsiveness. These principles will play a fundamental role in achieving the main objectives of the ITSs which include access and mobility, environmental sustainability, and economic development [3,4].

Given the advances in information technology and communication, the concept of a networked vehicle has received immense attention all over the world. A current trend is to provide vehicles and roads with capabilities to make the transportation infrastructure more secure, more efficient, and urban aware, and to make passengers' time on the road more enjoyable. In this context, a more secure transportation infrastructure means to provide information about traffic jams, accidents, hazardous road conditions, possible detours, weather conditions, and location of facilities (e.g., gas stations and restaurants) [5]; more efficient means an increased road network capacity, reduced congestion and pollution [6], shorter and more predictable journey times, lower vehicle operating costs, more efficient logistics, improved management and control of the road network, and increased efficiency of the public transport systems. Vehicles can also be used to collect, analyze, and share knowledge of an Area of Interest (AoI) in applications such as civilian surveillance (photo shots of violent scenes in progress sent to public authorities via infrastructure), pollution control, roads and traffic planning and innumerable others urban-aware applications. Finally, more enjoyable means to provide Internet access, tourist/advertising information, social media on the road, guidance for people to follow each other on the road, games, file downloads, and social applications (e.g., microblogs and chats) [7]. These applications are typical examples of what we call an ITS, whose goal is to improve safety, efficiency, urban awareness, and enjoyment in transportation systems through the use of new technologies for information and communication.

The creation of a Smart Road Environment (SRE) is an important direction in the Smart & Safe City concept. SRE is required for the interaction of satellite vehicle monitoring systems, ITS, unmanned vehicles, intelligent road infrastructure

 DOI: 10.1201/9781032691787-8

components, and mobile communication users [8]. The main elements of the SRE are (1) Intelligent real-time monitoring system, (2) Real-time traffic information system for alerting and warning of road users, (3) System of accounting and analysis of road users social reactions, (4) Interactive journey planner system, (5) Intelligent road lights systems, (6) Intelligent traffic signaling system, (7) Surveillance cameras (CCTV), photo-radar complexes, (8) Satellite systems of transport monitoring, (9) Parking and loading areas information systems, (10) Sensor systems for the movement of unmanned vehicles, (11) Intelligent vehicle transport systems, (12) Electronic payment systems, etc. An important element of SRE is a real-time controlling and monitoring system for decision-making on the management of road infrastructure objects. The system works with sensing systems such as a network of photo-radar vehicle detectors for road accidents, video surveillance cameras, vehicle information and communication systems (VICS), built-in car navigation equipment, and mobile communication equipment. It is designed for sensor data collection and processing. The monitoring objectives are analysis, assessment, and forecast of changes in traffic situations to control the behavior of vehicles and road users and alert police, emergency services, ambulance, maintenance, and other services.

8.2 SENSING FOR IMPROVEMENT OF TRANSPORTATION SYSTEMS

To improve traffic conditions, there needs to be effective sensing of the environment, and this is where sensor systems play an integral role. These systems—such as passive infrared and daytime traffic cameras or active radar systems—can gather vast amounts of data. To cope with this, AI-driven algorithms are increasingly being used to alert road operators to issues as well as to fuse data from numerous sensor sources for a richer information environment [3]. This wealth of information from traffic sensors can also be used for predictive analysis to improve traffic flow and other infrastructure challenges that can arise over time. The advances in sensing and communication provide tremendous opportunities to create smart transportation systems that optimally utilize the roadway infrastructure to minimize travel time, improve safety, and reduce emissions. Developments in technologies present Road Administrations around the world with the opportunity to transform the way that they manage and operate their highway networks.

The success of ITS depends on the sensing platform used to gather, process, and analyze data from the roadway and vehicular environment. Sensing platforms are composed of various types of sensing instruments mounted on roadway infrastructure and vehicles [1]. The first category is the intravehicular sensing platform which collects data about a vehicle's conditions. The second category, urban sensing platforms, is used to collect information about traffic conditions [1]. They are responsible for gathering the vehicular environment's physical variables such as traffic flow, speed, density, and occupancy. The platform can specify the setting of each sensor (such as the type of physical variable to sense, periodicity of sensing, and coverage, among others). Sensor technology is a vital component used for data collection during Vehicle to Vehicle (V2V) and Vehicle to Infrastructure (V2I)

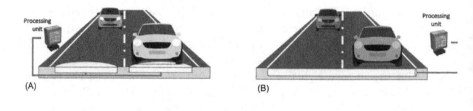

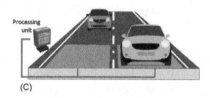

FIGURE 8.1   Intrusive sensors. (a) Magnetic sensors. (b) Pneumatic tube. (c) Inductive loops.

communications. These sensors form a network similar to the Internet of Things (IoT) that can gather data from different sensors (cameras, radars, loop detectors, among others) to be analyzed in near real-time to improve road traffic conditions (congestion, traffic flow, among others). This data is then provided to transportation management systems for further processing and analysis and subsequent decisions/ actions. Once the transportation management center receives the collected information, it is further processed and analyzed to make subsequent decisions/actions to reduce traffic congestion and create safer roads. See Figure 8.1 for the illustration of sensing in modern ITS.

A broad range of traffic sensors is deployed inroads for measuring physical traffic variables. These sensors can be categorized based on many criteria; one of the main criteria is whether the sensor is stationary or moving with the traffic flow. From this perspective, traffic sensors are generally categorized into two classes: (1) stationary sensors and (2) mobile sensors. Stationary sensors are installed at a fixed location near or on the roadways, whereas the latter sensors move with the vehicle stream. Traditionally, traffic sensing has been performed mainly utilizing stationary sensors. Although the paradigm has shifted with the emergence of mobile sensors such as probe vehicles and connected vehicles (CVs), fixed sensors still play a significant role in traffic sensing. There exist challenges and shortcomings in both sensing approaches. Data integration (fusion) is a potential solution to establish a robust sensing mechanism. Stationary and mobile sensors are discussed in detail in the following two sections.

8.3   SENSOR TECHNOLOGY AND COMMUNICATION ADVANCES

Over the last decade, sensor technology has become ubiquitous and has attracted a lot of attention. Sensors have been deployed in many areas such as forest [9,10], vehicle, and marine [11,12] monitoring. In transportation, sensor technology supports the design and development of a wide range of applications for traffic control, safety, and entertainment. In recent years, sensors, and actuators such as tire-pressure sensors and rear-view visibility systems have become mandatory (due to federal regulation in the United States [13]) in the manufacturing of vehicles and the implementation of ITSs, aimed at providing services to increase drivers' and passengers' satisfaction, improve road safety and reduce traffic congestion. Other sensors are optionally installed by manufacturers to monitor the performance and status of the vehicle and provide higher efficiency and assistance for drivers. Currently, the average number of sensors in a vehicle is around 60–100, but as vehicles become "smarter", the number of sensors might reach as many as 200 sensors per vehicle [14]. In Ref. [15], the author presents a classification of three categories of sensors based on the place of deployment in the vehicle: powertrain, chassis, and body. Another work classifies sensors in a vehicle based on the type of application the sensor is intended to support, and four categories of sensors are identified: sensors for safety, sensors for diagnostics, sensors for convenience, and sensors for environment monitoring [16]. Other extend the classification (four categories) proposed in Ref. [16] to include two additional categories of sensors, namely sensors for driving monitoring and traffic monitoring, as shown in Table 8.1.

TABLE 8.1
Categories of Sensors

Category of Sensors	Description	Example
Safety	Form the basis of safety systems and focus on recognizing accident hazards and events almost in real-time.	Micromechanical oscillators, speed sensors, cameras, radars and laser beams, inertial sensors, ultrasonic sensors, proximity sensors, night vision sensors, haptic.
Diagnostic	Focus on gathering data for providing real-time information about status and performance of the vehicle for detecting any malfunction of the vehicle.	Position sensor, chemical sensors, temperature sensors, gas composition sensors, pressure sensor, airbag sensor.
Traffic	Monitor the traffic conditions in specific zones, gathering data that improves the traffic management.	Cameras, radars, ultrasonic, proximity.
Assistance	Responsible for gathering data that provide support for comfort and convenience applications.	Gas composition sensor, humidity sensors, temperature sensors, position sensors, torque sensors, image sensors, rain sensors, fogging prevention sensors, distance sensors.
Environment	Monitor the environment conditions, offering drivers and passengers alert and warning services that are used to enhance their trips.	Pressure sensors, temperature sensors, distance sensors, cameras, weather conditions.
User	Focus on gathering data that support the detection of abnormal health conditions and behavior of the driver that can deteriorate the driver's performance.	Cameras, thermistors, Electrocardiogram (ECG) sensors, Electroencephalogram (EEG) sensors, heart rate sensor.

8.4 AUTOMOTIVE SENSORS

In recent years, the work on autonomous vehicles has increased owing to the need to automate one of the most human-intensive tasks on the road: Driving. 88%–95% of road accidents are due to human error [1]. There is very little margin of error on the software end available since running autonomous cars on the road in challenging conditions is life-critical. Human errors could be due to multiple problems: recognition error, decision error, performance error, or nonperformance (sleep, etc.) error. Bringing complete autonomy to the road is one of the present day's toughest challenges. Complete autonomy has multiple benefits, fewer accidents and loss of human lives, more time for humans to do other tasks, fleet optimization for reduced energy consumption, time saving (by optimized movement of vehicles on the road, etc.) and so on. Extensive research has been carried out in the industry to solve the challenges of environmental perception. This involves bridging the gap between computer-based decision-making and human-based decision-making. Understanding the environment requires the use of different sensors, and a complex autonomous system requires the use of multiple sensors working synchronously providing sufficient environmental data. These sensors are not new in the industry.

For autonomous driving, combinations involve sensors previously used in the automotive industry, like global positioning sensor (GPS), inertial measurement unit (IMU), etc., and perception sensors for autonomous driving, like camera, light detection and ranging (LiDAR), Radar, etc. Using multiple sensors is accompanied by heavy on-board computational requirements, and how to effectively integrate and process the data in a real-time application is in the list of challenges.

Perception sensors are responsible for accessing the environment in front and around the vehicle. There are two types of perception sensors, active, e.g. camera, and passive, e.g. LiDAR. An active sensor is a sensing device that requires an external source of power to operate; active sensors contrast with passive sensors, which simply detect and respond to some type of input from the physical environment. Although these sensor types can be used independently depending on application; for autonomous driving, using a combination of active and passive sensors ensures that one takes care of the demerits of the other. Data Fusion is a complex task and involves various difficulties like latency of the sensors, time-stamping the data, obtaining a consistent sensor data stream, etc. This happens because all the sensors have different data transmission rates (frequency) and the amount of data collected per time frame is different for all sensors. There are different data fusion techniques, depending on computational complexity and required amount of data required to assess the situation. Some of the techniques of multiple sensor data fusion will be discussed further here. Testing the algorithms on a real vehicle has its own set of challenges, and involves legal permissions to run it on public roads for testing. Moreover, it is still a threat to human lives, since testing requires a safe environment. Implementation of algorithms on a scaled vehicle is a good way to test the algorithms, and virtual environments provide another way to validate algorithms.

8.5 STATIONARY SENSORS (IN ROAD SENSORS)

Even though the automotive industry has invested a lot of money to increase safety, performance, and comfort in vehicles using sensors within the vehicle; traffic data collection using mechanisms located along the roadside has become one of the main challenges for ITSs. Sensor deployment within a transportation network provides drivers with new services such as smart parking (e.g., matching drivers with available parking spots) and reduced pricing according to congestion levels on the road. Sensors collect environmental data in real-time which is then processed and analyzed to improve transportation networks and make them resilient. Traditionally, stationary or fixed location sensors have been used to sense vehicular movement on roadways. These sensors can capture the traffic situation of a particular space at different temporal frames. Hence, data obtained from fixed location sensors are in the Eulerian reference frame [17]. Sensing using stationary sensors is also referred to as infrastructure sensing as the sensors are placed on the roadway infrastructure. Fixed sensors offer excellent temporal resolution as they detect vehicles at frequent time intervals. However, these sensors can cover a limited geographical area. As a result, the spatial resolution of the data obtained from the stationary sensors depends on the proximity of adjacent sensors. Moreover, installing fixed sensors is often costly, and their maintenance is costly too, which is a significant drawback. According to their locations on the roadway infrastructure, these sensors are further categorized into two classes: intrusive and nonintrusive [18]. Intrusive sensors are installed on pavement surfaces. They have high accuracy, but they also have high installation and maintenance costs. Basically, intrusive sensors (as shown in Figure 8.1) can be classified into three groups: (1) passive magnetic sensors which are installed on roads and are connected either wired or wirelessly to processing units (2) pneumatic tube sensors placed across the road which transmit data to processing units through wired/ wireless media, (3) inductive loops that are wire coils buried into roads and send data to processing units. This group of sensors is the most used in traffic control systems [19]. A short description of the characteristics of different stationary sensors is presented here.

8.5.1 INTRUSIVE SENSORS

Intrusive sensors are commonly deployed on pavement surfaces and have high accuracy, but installing these sensors is very expensive, and their maintenance costs are also high. Basically, intrusive sensors (as shown in Figure 8.2) can be classified into three groups: (1) passive magnetic sensors which are installed on roads and are connected either wired or wirelessly to processing units, (2) pneumatic tube sensors placed across the road which transmit data to processing units through wired/wireless media, (3) inductive loops that are wire coils buried into roads and send data to processing units. Passive magnetic sensors are mounted within holes in the roads and are connected to the processing units (Figure 8.2a). Magnetic sensors are passive devices that can detect vehicles by measuring the perturbation or magnetic anomaly caused in the Earth's magnetic field due to the presence of a vehicle (ferrous metallic object). Generally, each sensor has a zone of detection (yellow zone).

The advantages of passive magnetic sensors are that they can be used with minimum intrusion and at certain locations, such as bridge decks, where inductive loops

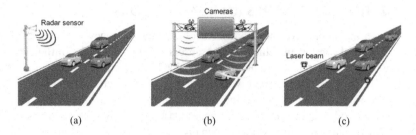

(a)   (b)   (c)

FIGURE 8.2 Nonintrusive sensor groups. (a) Roadside mast-mounted. (b) Bridge mounted. (c) Across roadside.

are not practical. These sensors provide more robust performance in unfavorable weather conditions, for example, snowfall and rain. They also show more tolerance to traffic stresses than the inductive loops. Pneumatic tube sensors are placed on the road and transmit data to processing units through wired/wireless media (Figure 8.1b). The sensor can sense the sudden change in air pressure when a vehicle tire passes over the tube. Pneumatic tube sensor's building material is fragile and easily damaged by heavy vehicles, so their usage is only temporal. Their main advantages are low cost, quick installation, and easy maintenance for temporary data recording. The limitations are the inaccurate axle counting when the volume of big vehicles is high and the fragility of the tubes.

Inductive loop detector is composed of wire coils embedded into roads (Figure 8.1c). Inductive loops are among the most commonly utilized traffic sensors due to their low cost, mature technology, wide deployment, and high accuracy in detecting vehicles. They were introduced in the early 1960s. Loop detectors with superior computation power were available by the 2000s [20]. The loop detectors can provide not only the data on vehicle count but also on vehicle speed, occupancy, headway, and gap [21]. Its technology is well understood, and its performance remains unaffected in adverse environmental conditions. It provides better accuracy for traffic count data than other commonly used sensors and is also insensitive to inclement weather conditions.

However, obtaining high spatial-resolution data from a loop detector can be very expensive. Inductive loops have high installation costs, suffer from radio interference if they are too close to each other, can be damaged during street maintenance, and can cause traffic disruption during installation, maintenance, and repairs. So, it can be concluded that the main advantage of road sensors is their technology maturity. They have been widely implemented and have high accuracy in detecting vehicles. However, the main disadvantages of road sensors are high installation costs, and traffic disruption during installation, maintenance, and repairs. One solution that has been implemented to address the aforementioned drawback is the introduction of wireless battery-powered sensor nodes which replace the intrusive sensors and are installed over the pavement. This technology represents a change in the transportation sensors which are expected to improve the quality, quantity, accuracy, and trustworthiness of the data collected from roads and avenues at a lower cost than current solutions [22–24].

8.5.2 Nonintrusive Sensors

Nonintrusive sensors cause minimal disruption to normal traffic operations during the process of installation, operation, and maintenance. Nonintrusive sensors are installed at different places on the roads (other than over it) as shown in Figure 8.2 and could detect a vehicle's transit and other parameters such as vehicle speed, and lane coverage. However, they are expensive and may be affected by environmental conditions. Normally, nonintrusive sensors are used to develop applications that provide information on a selected location, such as queue detection at a traffic light, traffic conditions, weather conditions of the road, and the pavement. Some sensors are mounted on a mast and are used to monitor a specific coverage area. Other sensors are mounted on bridges with a monitoring area directly below. Finally, some sensors are placed road side at ground level and use a beam that crosses the road and are mainly used for a single lane and with unidirectional flows because they are very susceptible to interferences from other objects.

Microwave radar sensors are widely used for sensing traffic conditions. In the United States, the allocated frequency for microwave radars is 10.525 GHz. Microwave radars are very compact, and they can detect vehicles from multiple lanes, which loop detectors cannot. Radar sensors are Doppler microwave devices that transmit low-energy microwave radiation reflected by all objects within the detection zone (Figure 8.2a). Radar sensors use the frequency shift of the reflected signal to detect moving vehicles and determine the vehicle's speed. In general, radar sensors are accurate and easy to install. They support multiple detection zones and can operate during the day or night. Their main disadvantage is high susceptibility to electromagnetic interference, and they cannot detect the vehicles driven slower than a cutoff speed. Laser sensors are active infrared sensors that transmit laser beams and measure the returning time of the reflected signal (Figure 8.2b). If an object is in the path of the laser beam, the returning time will be from an object. When a vehicle crosses the laser beam, the signal return time is reduced, and the car is detected. These sensors work well all day and are not affected by sun glare, headlight glare, or shadows. Laser sensors can detect traffic volume, the presence of a vehicle, its length, speed, and the number of axles. However, the sensors can be affected by occlusion, dirt, snow, and road grime on the lens.

Photo-radar complexes for data collection: Monitoring of objects and incidents in the road infrastructure is carried out by collecting and processing sensor data obtained from ground platforms, and aerial and space surveillance facilities. A ground platform in SRE are photo-radar vehicle detector complexes (Figure 8.3). Photo-radar complexes allow an automatic mode to fix incidents on objects of a road-transport infrastructure, to collect and accumulate sensor data [25]. A lot of complexes receive a huge amount of data, which cannot be processed by a person in real-time. Complexes can recognize objects in photos and a video stream, measure the speed of vehicles in the control zone, automatically capture and save photos of violators, recognize license plates, and collect and transfer data to the data center. However, the complexes do not have the capabilities of intellectual analysis and forecasting in real-time mode.

Inductive loops: Inductive loops can be placed in a roadbed to detect vehicles as they pass through the loop's magnetic field. The simplest detectors simply count the

FIGURE 8.3   CCTV cameras and photo-radar vehicle detector complexes [25].

number of vehicles during a unit of time (typically 60 seconds in the United States) that pass over the loop, while more sophisticated sensors estimate the speed, length, and class of vehicles and the distance between them. Loops can be placed in a single lane or across multiple lanes, and they work with very slow or stopped vehicles as well as vehicles moving at high speed.

Bluetooth: Bluetooth is an accurate and inexpensive way to transmit position from a vehicle in motion. Bluetooth devices in passing vehicles are detected by sensing devices along the road. If these sensors are interconnected, they are able to calculate travel time and provide data for origin and destination matrices. Compared to other traffic measurement technologies, Bluetooth measurement has some differences:

- Accurate measurement points with absolute confirmation to provide to the second travel times.
- Is nonintrusive, which can lead to lower-cost installations for both permanent and temporary sites.
- Is limited to how many Bluetooth devices are broadcasting in a vehicle so counting and other applications are limited.
- Systems are generally quick to set up with little to no calibration needed.

Since Bluetooth devices become more prevalent on-board vehicles and with more portable electronics broadcasting, the amount of data collected over time becomes more accurate and valuable for travel time and estimation purposes. Bluetooth detectors have recently become very popular due to the ubiquitous presence of Bluetooth technology in smartphones and vehicles. When successive roadside Bluetooth detectors detect the same car, important information on its speed and travel time can be computed. Apart from vehicular traffic, they are also useful in estimating the origin-destination (OD) matrix for vehicles and public transit passengers.

Traffic flow measurement and automatic incident detection using video cameras is another form of vehicle detection. Since video detection systems such as those used in automatic number plate recognition do not involve installing any components directly into the road surface or roadbed, this type of system is known as a "nonintrusive" method of traffic detection. Video from cameras is fed into processors that analyze the changing characteristics of the video image as vehicles pass. The cameras are typically mounted on poles or structures above or adjacent to the roadway. Most video detection systems require some initial configuration to "teach" the processor the baseline background image. This usually involves inputting known measurements such as the distance between lane lines or the height of the camera above the roadway. A single video detection processor can detect traffic simultaneously from one to eight cameras, depending on the brand and model. The typical output from a video detection system is lane-by-lane vehicle speeds, counts, and lane occupancy readings. Some systems provide additional outputs including gap, headway, stopped-vehicle detection, and wrong-way vehicle alarms.

Video detection systems include video cameras, a computer for data processing, and sophisticated algorithms used to interpret the images and translate them into traffic data (Figure 8.2c). They can monitor multiple lanes and multiple detection zones simultaneously. Along with the usual traffic parameters (flow and speed), the raw video data can provide rich insights such as near misses, the composition of vehicles on the road, and the causes of an incident. The main disadvantage of video detection systems is their susceptibility to poor performance in bad weather conditions. Moreover, it is susceptible to camera motion or vibration of mounting poles due to winds.

Nonintrusive sensors: Nonintrusive sensors provide many of the intrusive sensors' functions with fewer difficulties. However, they are highly affected by climate conditions such as snow, rain, and fog, among others. Accurate traffic data is of utmost importance to make informed decisions to improve traffic conditions. Nonintrusive sensors are more easily spotted by drivers, resulting in different and faster reactions such as: slowing down and using the correct drive lane, among others after detecting those devices. The challenge is not just the installation of these sensors, but also reducing the drivers' reaction times based on the collected data and providing them with a more precise view of the context and the reality of the road or avenue. Currently, several sensors are used on roads. Table 8.2 shows the two categories (intrusive and nonintrusive) of sensors that are currently used for keeping track of the number of vehicles, vehicle classification, or road conditions [26] as well as some other practical uses.

Pneumatic road tube sensors use one or several tubes placed across traffic lanes allowing for number of vehicles counting and vehicle's classification. When a vehicle's

TABLE 8.2
Two Categories (Intrusive and Nonintrusive) of Sensors

Category	Sensor Type	Application and Use
Intrusive	Pneumatic road tube.	Used for keeping track of the number of vehicles, vehicle classification and vehicle count.
	Inductive Loop Detector (ILD).	Used for detection vehicle's movement, presence, count and occupancy. The signals generated are recorded in a device at the roadside.
	Magnetic sensors.	Used for detection of presence of vehicle, identifying stopped and moving vehicles.
	Piezoelectric.	Classification of vehicles, count vehicles and measuring vehicle's weight and speed.
Nonintrusive	Video cameras.	Detection of vehicles across several lanes and can classify vehicles by their length and report vehicle presence, flow rate, occupancy, and speed for each class.
	Radar sensors.	Vehicular volume and speed measurement, detection of direction of motion of vehicle and used by applications for managing traffic lights.
	Infrared.	Application for speed measurement, vehicle length, volume, and lane occupancy.
	Ultrasonic.	Tracking the number of vehicles, vehicle's presence, and occupancy.
	Acoustic array sensors	Used in the development of applications for measuring vehicle's passage, presence, and speed.
	Road surface condition sensors	Used to collect information on weather conditions such as the surface temperature, dew point, water film height, the road conditions and grip.
	RFID (Radio-frequency identification)	Used to track vehicles mainly for toll management.

tire passes over the tube, the sensor sends a burst of air pressure which produces an electrical signal. The electrical signal is transmitted to the processing unit [26].

The Inductive Loop Detector (ILD) sensor is one of the most common sensors in traffic management. It is used for collecting traffic flow, vehicle's occupancy, length, and speed. It consists of a long wire coiled to form a loop which is installed into or under the surface of the road and measures the change in the electrical properties of the circuit when a vehicle passes over the sensor, producing an electrical current that is sent to the processing unit.

Magnetic sensors are used to detect vehicles when a change in the Earth's magnetic field is produced. Magnetic sensors are used to collect flow, occupancy, vehicle length, and speed and are suitable for deployment on bridges.

Piezoelectric sensors detect vehicles passing over (at high-speed ranges around 112 km/h) a sensor through a change in the sensor's voltage and can monitor up to four lanes. Piezoelectric systems are commonly formed by piezoelectric sensors and ILD sensors.

A Video Image Processor (VIP) system includes several video cameras, a computer for processing the images, and sophisticated algorithm-based software for interpreting the images and translating them into traffic data. Video cameras placed at the roadside collect and analyze video images from a traffic scene to determine the changes among successive frames using traffic parameters such as flow volume and occupancy. The main disadvantage of VIP systems is that they are susceptible to reduced performance caused by bad weather conditions.

Radar sensors transmit low-energy microwave radiation that is reflected by all objects within the detection zone. There are different types of radar sensor systems: (1) Doppler systems that use the frequency shift of the return to track the number of vehicles, and determine speed very accurately, (2) frequency-modulated continuous wave radar radiates continuous transmission power such as a simple continuous wave radar and is used to measure flow volume, speed, and presence. In general, radar sensors are very accurate and easy to install. They support multiple detection zones and can operate during the day or night. Their main disadvantage is high susceptibility to electromagnetic interference.

Infrared sensors detect the energy generated by vehicles, road surfaces, or other objects. Basically, sensors convert the reflected energy into electrical signals that are sent to the processing unit. Infrared sensors are divided into two categories: Passive Infrared (PIR) detects vehicles based on emission or reflection of infrared radiation and is used to collect data from flow volume, vehicle presence, and occupancy. Active InfraRed (AIR) sensors use Light-Emitting Diodes (LED) or laser diodes to measure the reflection time and collect data on flow volume, speed, classification, vehicle presence, and traffic density.

Ultrasonic sensors calculate the distance between two objects based on the elapsed time between a sound wave transmitted at frequencies between 25 and 50 kHz and reflected to the sensor by an object. The received energy is converted into electrical energy which is sent to the processing unit. Ultrasonic sensors are used to collect data about vehicle flow and the vehicle's speed. The main disadvantage of this kind of sensor is its high sensitivity to environmental effects.

Acoustic array sensors are formed by a set of microphones that are used to detect an increase in sound energy, produced by an approaching vehicle passing through the coverage area of the sensor. Acoustic sensors are replacing magnetic induction loops to calculate traffic volume, occupancy, and average speed of vehicles.

Road surface condition sensors use laser and infrared technologies to read road conditions (temperature and grip) to improve traffic safety and execute road maintenance programs. However, this type of sensor requires periodic maintenance to maintain its performance level.

Radio-Frequency ID (RFID) sensors are used for (1) automatically identifying running vehicles on roads and collecting their data, (2) for smart parking, and for detecting vehicles to allocate space for parking.

Even though many sensors have been installed on roads and streets, the lack of correct calibration and cluster integration makes the data collected unstable and hinders the development and evolution of ITS as projected and expected from transportation authorities, car makers, road users, and all ITS stakeholders [26]. ITS are expected to use all kinds of integrated sensors to provide situation evaluation systems, and fast decision-making based on the data collected from integrated sensors to improve transportation conditions.

8.6 MOBILE SENSORS

A mobile sensor observes a traffic environment from a dynamic reference frame. The mobile sensors (usually deployed on vehicles) move in traffic streams providing the measurements in the Lagrangian reference frame [27]. Moving sensors can complement the low-spatial resolution of fixed location sensors as they collect traffic data in the unobservable areas between two fixed sensors. The necessity of moving sensors has always been there, but it was not technologically feasible. However, now its application is widespread with the introduction of connected and autonomous vehicles (CAVs) and smartphones. A short description of various mobile traffic sensors is given here.

8.6.1 PROBE VEHICLES

The most prominent examples of mobile traffic sensors are probe vehicles or floating cars. The idea of a probe vehicle was introduced in Japan in the early 1970s. However, the concept was not widespread up until the revolution of wireless communication technology. Initially, researchers drove a vehicle on the route of interest and recorded the vehicle position and time at regular time intervals termed probe data measurements. Thus, probe vehicles serve as sources of vehicle trajectory and travel time data. Probe vehicles nowadays use various wireless technologies such as Bluetooth and cellular networks for data transmission making the data collection real-time. Probe vehicles offer wide spatial coverage. However, probe vehicles have some disadvantages. Probe vehicle data can be highly biased as taxis, buses, and commercial vehicles are often used as probe vehicles. For example, in Beijing, 10,000 taxis equipped with GPS were introduced as probe vehicles in the early 2000s. Taxis may use different route choices, and buses may have dedicated routes. Thus, depending solely on probe vehicle data for travel time estimation can be misleading. Other examples, Probe-Vehicle—In some countries, taxis/government-owned vehicles equipped with dedicated short-range communication (DSRC) or other wireless technology are often used as probe vehicles to send the probe data such as their locations and speed, to a centrally operated traffic management center (TMC). The TMC then aggregates the received data to get an impression about the traffic flow and congested locations of an area.

8.6.2 CONNECTED VEHICLES

A CV is a vehicle that can communicate bidirectionally with the infrastructure or other systems outside of the vehicle. CVs can communicate with roadside infrastructures and provide their speed, location, and trajectory information. Thus, CVs can be seen as probe vehicles that trace the infrastructure on a real-time basis, and this information helps speculate the current traffic situation. CVs can generate and deliver a massive volume of traffic data about vehicle trajectory and density. Researchers are anticipating that CVs would function as a potent tool for real-time traffic state estimation. As the market penetration rate of CVs increases, their influence on traffic state estimation might grow significantly.

8.6.3 Autonomous Vehicles

An autonomous vehicle (AV) or self-driving car is a vehicle that can sense the road environment and can drive itself safely with little or zero input from a human driver. An AV is equipped with many sensors, including radar, LiDAR, wide-angle cameras, GPS, etc. These sensors are vital for AV's perception and localization purpose. At the same time, they also provide valuable knowledge about surrounding traffic conditions. For example, a LiDAR can accurately map its surroundings during both day and night. Short- and long-range radars can help detect nearby vehicles and pedestrians. The fusion of LiDAR and radar data will provide accurate vision to the AVs even in extreme weather conditions. Most of the AVs are anticipated to be also equipped with CV technology. Thus, AVs cannot only provide their own trajectory data (like CVs) but also provide nearby traffic information such as traffic density, speed, any incidents, fog or snow conditions, etc. This will give an unprecedented superiority in sensing over conventional stationary detectors. However, the penetration rate of AVs is still meager to make any significant impact on road network sensing.

8.6.4 Unmanned Aerial Vehicles

An unmanned aerial vehicle (UAV) can fly without a human piloting it. It may be autonomous or remotely controlled from a ground station. UAV technology has rapidly advanced in the last few years, and one of the leading applications includes its use in remote sensing, among many others. The application of a video camera-equipped UAV as a traffic sensor is unique because it is not located at the road level, unlike other traffic sensors. Thus, a UAV has the advantage of observing a broader traffic network. Moreover, the UAV is neither stationary nor moves with the traffic flow. That means the observations are different than what we observe from static sensors (Eulerian frame) and the in-stream moving sensors (Lagrangian frame). Fusion of data obtained from UAV and other fixed/moving sensors can be of immense use to estimate traffic flow characteristics. Also, Radio wave or infrared beacons are used as in Japan to transfer real-time traffic information arterial roadways, a moderate capacity roadway just below highways in level of service.

8.7 INTELLIGENT VEHICLE SENSING

This section furnishes additional details on intelligent vehicles and their potential use in traffic sensing. A generally accepted definition of an intelligent vehicle is a vehicle that operates at a certain level of autonomy and assists the driver in performing specific driving tasks more effectively—thereby resulting in improved safety, fuel efficiency, and reduced travel times. CAV is another commonly used term often used in the same context. In ITS, identifying the type of sensors to develop applications that contribute to address problems such as: (1) traffic congestion and parking difficulties, (2) longer commuting times, (3) higher levels of CO_2 emissions, and (4) increase in the number of road accidents, among others is of critical importance for improving a vehicle's performance as well enhancing the driving experience.

8.7.1 Connected and Autonomous Vehicles

The emergence of CAVs promises a fundamental change in how people travel and interact with vehicles. CVs have the ability to communicate with other neighboring vehicles, infrastructure, pedestrians, and with everything technology enabled. This connectivity enables a vehicle to be a part of a larger IoT type network, share and receive information about road conditions, live traffic conditions, movement of emergency vehicles, signal timing/phases, etc. An AV does not require a driver. While the complete autonomy of vehicles is still in the development and testing phase, partially autonomous vehicles with various automation levels are available in the market. The partial automation includes automatic parking, advanced driver assistance, emergency braking system, lane control, adaptive cruise control, etc. Although connected and automated vehicle technologies are not interdependent, they can provide safer, efficient, and more reliable commute options when combined.

8.7.2 Sensors in Vehicles

As discussed earlier, sensors and actuators are becoming primordial elements in newer vehicles as shown in Figure 8.4. A wide range of sensors such as LiDAR, radar (radio detection and ranging), stereo camera, global navigation satellite system (GNSS), IMU, and ultrasonic distance sensors are used in AVs. Currently, a vehicle has an average of around 60–100 sensors installed, but as vehicles become "smarter," the number of sensors might reach as many as 200 sensors per vehicle. Sensors can be classified based on their deployment in the vehicle and the type of applications

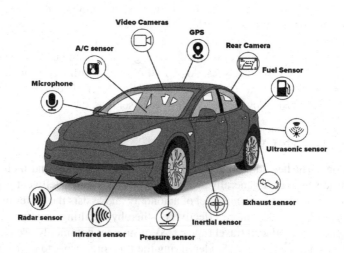

FIGURE 8.4 Sensors in autonomous vehicle.

TABLE 8.3
Sensors in Vehicle

Category	Sensors	Description
Safety	Micromechanical oscillators, speed sensors, cameras, radars, laser beams, inertial sensors, ultrasonic sensors, proximity sensors, night vision sensors, haptic.	Form the basis of safety systems and focus on recognizing accident hazards and events almost in real-time.
Diagnostic	Position sensors, chemical sensors temperature sensors, gas composition sensors, pressure sensors, airbag sensors.	Focus on gathering data for providing real-time information about the status and performance of the vehicle for detecting any malfunction of the vehicle.
Traffic	Cameras, radars, ultrasonic proximity.	Monitor the traffic conditions in specific zones, gathering data that improve traffic management.
Assistance	Gas composition sensors, humidity sensors, temperature sensors, position sensors, torque sensors, image sensors, rain sensors, fogging prevention sensors, distance sensors.	Responsible for gathering data that provide support for comfort and convenience applications.
User	Cameras, thermistors, electrocardiogram (ECG) sensors, electroencephalogram (EEG). Sensors, heart rate sensor.	Focus on gathering data that supports the detection of abnormal health conditions and the driver's behavior that can deteriorate the driver's performance.

they support. Table 8.3 shows the classification of vehicle sensors [28–30]. Figure 8.1 depicts some of the most widely used sensors in vehicles today.

Vehicle sensors have multiple advantages, and on-board information display makes it much easier for the driver to optimize resource and have a comfortable drive. Some of the sensors used in the automotive industry are temperature sensors for the engine, wheel speed sensors, humidity sensor, inertial measurement sensor (IMU), GPS, brake and type pressure sensor (TPS/BPS), collision sensor, etc. Most of the driver assistance systems use the data from these sensors to optimize the drive. For example, the Traction Control and Anti-Lock braking system are driver assistance system use the data from wheel speed and inertial measurement (IMU) sensor to avoid slipping of tires in challenging terrains and weather conditions. Further data acquisition from these sensors help vehicle manufacturers in subsequent design assessment, monitoring, maintenance, improving efficiency, etc.

8.7.3 APPLICATIONS FOR IN-VEHICLE SENSORS

Tire-pressure monitoring is an application that is required for the National Highway Traffic Administration of the U.S. to alert drivers using acoustic, light or vibration warning if the tire air pressure is low [31]. Proximity, ultrasonic and electromagnetic sensors are used in parking assistance and reverse warning applications. Proximity sensors can detect when a vehicle gets close to an object. Ultrasonic sensors use a type of sonar to identify how far the vehicle is from an object, alerting the driver when the vehicle gets closer than a set threshold. Electromagnetic sensors alert the driver when an object enters an electromagnetic field created around the front and back bumpers. Proximity sensors have been used to develop a system based on a rectangular capacitive proximity-sensing array for occupant head position quantification to meet the guidelines of the Insurance Institute for Highway Safety (IIHS) [32]. However, these types of sensors are frequently affected by temperature and humidity, reducing their accuracy.

RAdio Detection and Ranging (Radar) and laser sensors constantly scan the road for frontal, side and rear collisions and allow safety applications to adjust throttle and activate brakes to prevent potential collisions or risk situations by using radio waves to determine the distance between obstacles and the sensor. The application notifies the driver if something close to the vehicle is detected and automatically activates the brakes to avoid a collision. The gyroscope and accelerometer sensors are used in Inertial Navigation Systems (INS) to determine the vehicle's parameters such as vehicle position, orientation, and velocity. INS are used in conjunction with GPS to improve accuracy. Radar and speed sensors are used in applications that warn the driver of potential danger if changing lanes or wandering out of the lane is detected. The driver is usually warned through vibration in the seat or steering wheel or acoustically using an alarm.

Cameras are used to: (1) monitor the driver's body posture, head position and eye activity to detect abnormal conditions such as signs of fatigue or the vehicle behaving erratically (driving out of a straight line on the road or pedestrians crossing suddenly in front of the vehicle) and (2) execute night vision assistance applications to help drivers see farther down the road and detect objects such as animals, people or trees in the path that can cause a potential risky situation or an accident.

LiDAR has become in a key component for the evolution of autonomous vehicles. LiDAR enables a self-driving car (or any robot) to observe the world with a few special characteristics such as continuous 360-degree visibility and highly accurate depth information. LiDAR sensors continually fire off beams of laser light, and then measure how long it takes for the light to return to the sensor.

Although more sensors are in each vehicle, their integration with other components and the lack of widely accepted standards among different brands is a huge drawback in their adoption. In contrast, current automated systems are limited in their capacities. For example, Volvo's city safety speed limit is 50 km/h or less to avoid collisions with other vehicles or hitting motorcycles or cyclists. A city safety system is based on a laser unit, so in darkness conditions, the it can only detect a vehicle if its headlights and taillights are on and are clearly visible [33]. On-board sensors are essential for vehicles to sense their environment and collect information about other nearby road users. Generally speaking, CAVs employ different sensors to perform two main perception tasks: (1) environmental perception and (2) self-localization. The first task is detecting and tracking vehicles and pedestrians, road surface detection, road sign detection, and lane detection. The second task is detecting the relative and absolute positions of the vehicle.

Most automotive manufacturers use three primary sensors: cameras, radars, and LiDARs. Cameras maintain an accurate 360-degree visual representation of their external environment to identify objects, obstacles, traffic lanes, and complement them with thermal imaging of objects such as pedestrians and animals.

Some applications such as lane recognition, semantic image segmentation, and object identification use the cameras to detect all the elements (such as traffic signals, obstacles, pedestrians, cyclists, among others) of the environment around the vehicle. Radar sensors determine, in real-time, the speed, proximity, and relative object size compared to the vehicle. These sensors employ radio frequency technology to identify objects, detect obstacles and blind spots, and track vehicles and pedestrians' speed. Radar sensors can only detect objects with low-spatial resolution, and they cannot distinguish among multiple entities. It is difficult for radar sensors to separate objects by their direction of motion. The detection of objects depends on the strength of radio wave reflection, which can be influenced by factors such as the object's size, absorption features, reflection angle, distance, and transmission power. LiDAR sensors work similar to radar systems, but LiDAR uses lasers instead of radio waves. LiDAR provides a vehicle with 360 degrees view of the surrounding environment and provides shapes and depth to surrounding objects. LiDAR helps vehicles to determine the relative vehicle's position according to other surrounding objects' locations.

Even though AV technology has come a long way in a short time, it still may take a few more years to witness the ubiquitous presence of AVs. Issues such as sensor failures in nonideal weather conditions—rain or snow—persist. The mass adoption of fully autonomous cars (i.e., level 5 AVs) is heavily dependent on how well and quickly the safety concerns are addressed. Much of the safety concerns can be alleviated by developing a reliable and fault-tolerant multimodal sensor fusion system in AVs as shown in Figure 8.5.

8.8 VEHICULAR COMMUNICATION SYSTEMS

Vehicular communication systems are computer networks in which vehicles and roadside units are the communicating nodes, providing each other with information, such as safety warnings and traffic information. They can be effective in avoiding accidents and traffic congestion. Both types of nodes are DSRCs devices. DSRC works in 5.9 GHz band with bandwidth of 75 MHz and approximate range of

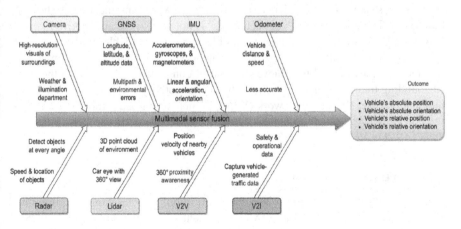

FIGURE 8.5 Various sensors in autonomous vehicles.

FIGURE 8.6   Vehicle to everything (V2X) communication.

300 m (980 ft) [34]. Vehicular communications is usually developed as a part of ITS. Communication between vehicles is evolving. It can be used for reducing dangerous situations, improving safety in traffic as well as making driving more convenient by enabling improved in-car entertainment such as Wifi and Video streaming.

8.8.1 Vehicle to Everything Communications

Various communication protocols and technologies have been defined to exchange data among the elements of the vehicular environment. At the top of vehicular communication systems is the vehicle to everything (V2X) communication as shown in Figure 8.6. The concept of a "connected car" is not new to the automotive industry, however, the technology to make it possible were not available until a few years ago. You can think of V2X as the parent category of a broader set of communication technologies needed to achieve the goal of connecting vehicles with the world surrounding them. The main types of vehicle connectivity encompassing V2X communications are: Vehicle to network (V2N); Vehicle to infrastructure (V2I); Vehicle to vehicle (V2V); Vehicle to cloud (V2C); Vehicle to pedestrian (V2P); Vehicle to device (V2D); Vehicle to grid (V2G).

8.8.2 Vehicle to Network

Thanks to vehicle to network (V2N) communication, vehicles can use cellular networks to communicate with the V2X management system. V2N also uses the DSRC standard to interact with other vehicles as well as the road infrastructure. This level of connectivity allows vehicles to be considered as a "device", just like smartphones, tablets, and wearables devices. Accessing mobile-network operators' LTE, 5G infrastructure and DSRC systems allow vehicles to:

- Receive broadcast alerts regarding road conditions (accidents, congestion, weather, etc.) also known as vehicle to infrastructure (V2I) direct communications.
- Communicate with nearby vehicles (via the cellular network and DSRC) also known as vehicle to vehicle (V2V) direct communications.
- Communicate with data centers and other devices connected to the Internet (vehicle to cloud communications).
- Establish communication with pedestrians' devices (vehicle to pedestrian communications).

Using LTE, 5G, and DSPC (which is considered an evolution of Wi-Fi exclusive to vehicles), V2N allows vehicles to reliably interact with infrastructure, other vehicles, other devices, and even pedestrians.

8.8.3 Vehicle to Infrastructure (V2I)

Vehicle to infrastructure (V2I) communication is an integral part of ITS that consists of bidirectional exchange of information between the vehicle and the road infrastructure. V2I allows data exchange between CAVs and transportation infrastructure, such as traffic signals, stop signs, and roadside units (RSUs). RSUs can communicate with a CAV within its communication range and collect vehicle data such as its position (latitude and longitude), heading angle, velocity, lateral acceleration, brake status, angle of steering, and turn signal status and OD. While at the same time, RSUs broadcast messages such as signal phase and timings (SPaT), basic safety messages (BSMs), and other messages containing road conditions and travel times. This information includes vehicle-generated traffic data gathered from other vehicles, data from sensors installed in the road infrastructure (cameras, traffic lights, lane markers streetlights, road signs, parking meters, etc.), and data broadcasted from the ITS (speed limits, weather conditions, accidents, etc.). As explained earlier, this information interchange is done wirelessly by means of the cellular network and/or DSRC frequencies. The technology behind V2I and ITS is endorsed by the European Union as well as the United States Department of Transportation ITSs Joint Program Office. The goal of V2I is to enhance road safety and prevent accidents by giving drivers real-time information regarding different conditions in the road. Moreover, V2I and ITS technologies are key to future autonomous vehicles that will rely on this valuable information. It was shown that Crossroads is very resilient to network delay of both V2I communication and Worst-case Execution time of the intersection manager [35].

On the other hand, for V2I communications, different access technologies have been proposed. Sensors such as passive magnetic sensors, pneumatic tube sensors, and inductive loops use 4G/LTE technology to disseminate data to vehicles, pedestrians, or traffic control offices. 4G/LTE offers a high data rate, low latency, and extensive coverage [36]. In recent years, technologies such as LTE-X2, Sigfox, and 5G have enabled many revolutionary applications. These technologies have become a viable solution for exchanging data among IoT devices, thus creating a global network to transmit data collected from different sensors located on roads and avenues [37]. The allocated 75 MHz spectrum to DSRC (IEEE 802.11p) in the 5.9 GHz band has remained underused, resulting in many automakers either abandoning the DSRC plan or switching to Cellular-V2X (vehicle to everything) technology.

8.8.4 VEHICLE TO VEHICLE (V2V)

To enable the envisaged smart mobility applications, interaction and information exchange among vehicles, road infrastructure, or on-road pedestrians is a must. As its name implies, vehicle to vehicle (V2V) communication enables vehicles to exchange data in real-time. This interchange is done wireless via DSRC frequencies, the same used in V2I communications. Thanks to V2V communication, vehicles can share their speed, location, position, or destination with other nearby vehicles as well as exchange any other relevant information, giving the system a 360-degree representation of its surroundings to make intelligent decisions, as shown in Figure 8.7. Since V2V communications were conceived as a mesh network, each vehicle becomes a node that can capture, send, and retransmit signals. Since V2V is an integral part of V2X and V2N, nodes also include smart traffic signals, road sensors, and other V2I components. According to NHTSA, V2V technology could prevent 615,000 motor accidents. As a result of the mesh design, vehicles fitted with V2V technology count with real-time information of everything happening in a radius of 300 m around them. Cutting-edge driving assist systems used by several top carmakers can use that information to alert the driver about an imminent hazard, instantly, enhancing road safety.

The automotive society has adopted communication technologies to exchange data among all elements of the transportation environment [38]. The access technologies implemented for V2V communications are IEEE Wireless Access in Vehicular

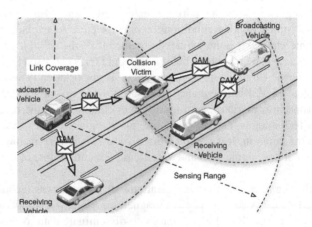

FIGURE 8.7 Communication modes for vehicles.

Environment (WAVE) standard, including the specification of DSRC, the IEEE 802.11p for PHY and MAC layers, and the IEEE 1609 family for upper layers [39]. Those technologies facilitate the data exchange over a range of 300m. Another standard used is the J2735, which specifies attributes such as messages, data frames, and data elements.

While both DSRC and C-V2X are radio access technologies using identical message payloads for use cases (V2V and V2I applications working the same way), the difference in their electronics forbids them to be interoperable. While DSRC has been available for many years now, the C-V2X has emerged and matured recently. Although at present, it is uncertain if the DSRC or C-V2X will win the battle, the emerging trends and automakers' stance suggests that C-V2X technology is here to stay (with major players like Ford and Qualcomm backing it). C-V2X uses the same technologies in 4G LTE (can support 5G) and can outperform DSRC in certain aspects. While C-V2X's communication capability with cellular networks has potential for vehicle to pedestrian (V2P), infrastructure to pedestrian (I2P), and in-vehicle infotainment applications—the radios can also directly communicate with each other. This direct communication, referred to as "side link," has the potential to match latency requirements for safety-critical BSMs. Large-scale pilots involving DSRC are already being tested in New York, Florida, and Wyoming, whereas C-V2X is yet to be tested in large-scale field deployment. Going forward, it is critical to have a clear understanding of emerging technological trends and the pros and cons of both DSRC and C-V2X technologies concerning the envisaged I2V and I2P applications. It can be concluded that V2V communication's ability to wirelessly exchange information about the speed and position of surrounding vehicles shows great promise in helping to avoid crashes, ease traffic congestion, and improve the environment. But the greatest benefits can only be achieved when all vehicles can communicate with each other. That's why NHTSA has been working with the automotive industry and academic institutions for more than a decade to advance V2V communication's lifesaving potential into reality.

8.8.5 Vehicle to Cloud (V2C)

Vehicle to cloud (V2C) communication leverages V2N access to broadband cellular mobile networks to offer data exchange with the cloud. Some applications of this technology include: Over the air (OTA) updates vehicles' software, Redundancy to DSRC communication, Remote vehicle diagnostics, Bidirectional communication with household appliances also connected to the cloud (IoT), Bidirectional communication with digital assistants. In a no distant future, V2C could also have an important role in shared mobility. For instance, drivers' preferences could be saved in the cloud and used when carsharing to adjust the seat position, mirrors, radio stations, and more automatically.

8.8.6 Vehicle to Pedestrian (V2P)

One of the newest subcategories of the V2X ecosystem is vehicle to pedestrian (V2P) communication. Other V2X technologies such as V2V and V2I involve cars and road infrastructure purposively prepared to communicate with each other. In some cases, like wheelchairs, bicycles, and strollers, a smart sensor could be implemented to create awareness about its presence. But pedestrians walking and children playing in the street are a whole different history. Some carmakers use in-vehicle systems such as

LiDAR technology to facilitate collision warnings, 360-degree cameras, and blind spot warnings to detect pedestrians. However, the reliability of such approaches varies. This explains why a new generation of handheld devices and mobile apps are being developed to make drivers aware of possible collisions.

8.8.7   VEHICLE TO DEVICE (V2D)

You can think of vehicle to device V2D as a subset of V2X communication that allow vehicles to exchange information with any smart device, usually via Bluetooth protocol. A typical application of this technology is Apple's CarPlay and Google's Android Auto that allow smartphones, tablets, and wearables to interact with the vehicle's infotainment system.

8.8.8   VEHICLE TO GRID (V2G)

Last but not least, vehicle to grid (V2G) communication is a relatively new member of the V2X group of technologies that provides bidirectional data exchange between plug-in hybrid vehicles (PHEV), battery electric vehicles (BEV), and even hydrogen fuel cell vehicles (HFCEV) with the smart grid in support of electrification of transport. Thanks to V2G communication, the next-gen electric grid will be able to balance loads more efficiently as well as reduce utility bill costs.

8.9   IOT PERSPECTIVE

Technological advances have transformed the vehicle into "a new mobile node" to connect all passengers, vehicles, and roadside infrastructure. As vehicles and passengers are connected, they can access, consume, create, and share digital content among pedestrians, passengers, vehicles, and traffic infrastructure, making the concept of the IoV [27]. The IoV comprises three fundamental elements:

1. The intravehicular network,
2. The intravehicular network, and
3. Vehicular mobile Internet.

These elements of IoV enable the information exchange among vehicles, road infrastructures, passengers, drivers, sensors, and actuators. For this purpose, IoV uses a set of communication protocols and standards such as IEEE802.11p, directional medium access control (DMAC), vehicular cooperative media access control (VC-MAC), ad hoc on-demand distance vector, dynamic source routing, general packet radio services, and others [40].

8.10   IOV ARCHITECTURE

The basic architecture model of the IoV contains three layers [41]. The first layer focuses on managing all sensors inside the vehicle in charge of data gathering and relevant event detection such as environmental conditions and vehicle conditions. The second layer controls the communication process. The layer supports different communication modes such as V2V, V2I, V2P, and vehicle-to-sensor. The central objective of this layer is to provide users with seamless connectivity using the existing and emerging communication networks. Finally, the last layer manages the intelligence of IoV, providing the support for storage, analysis, estimation, processing, and deciding on all the risk situations that the vehicle faces during the driving task.

Researchers have complemented the basic architecture with more layers. For example, CISCO proposed an IoV architecture based on four layers [42]. In Ref. [43], SAP proposed an architecture based on a service-oriented and event-driven architecture. This architecture includes two components: a back-end system for information exchange and a back-end manager for interconnection of the sensors and actuators deployed into the vehicle. Finally, Ref. [44] proposed a seven-layered model architecture for IoV that complements the previously mentioned architectures, providing a transparent interconnection of all elements and disseminating data into an IoV environment.

8.11 TRAFFIC STATE ESTIMATION

Active Transportation and Demand Management (ATDM) is the dynamic management and control of travel demand, traffic demand, and traffic flow using real-time traffic states variables such as aggregated traffic speed, volume, and occupancy. Traffic state estimation is the art of extracting traffic state variables from partially observed traffic variables [27,45–46]. The more formal definition of traffic state estimation can be given as follows: "The simultaneous estimation of flow, density, and speed on road segments with high spatiotemporal resolution based on partially observed traffic data and a priori knowledge of traffic" [47]. Traffic state estimation (TSE) is of paramount importance to traffic engineers. The ever-increasing demand for traffic safety, efficient management, and mobility has made real-time TSE a critical component of ITS. An accurate and robust TSE is critical as it helps traffic engineers to design efficient traffic control strategies [48]. The traffic states are collected primarily using fixed roadside sensors—leading to an Eulerian specification of the traffic flow field. Traffic sensors observe traffic conditions at a specific location or portion of a road segment. The overall traffic conditions of the segment can be reconstructed from partially observed real-time or historical data. Such sensors, however, have their limitations in terms of coverage, associated costs, resiliency, accuracy, and the measurements are often accompanied by noise. In worst cases, the sensor network can fail, for example, in a natural disaster or due to cyber-attacks. Therefore, the dependence on a particular sensor is not desirable. Integrating traffic data from multiple sources can make TSE more robust and fault-tolerant.

The process is known as heterogeneous data fusion. It increases the reliability of estimation by offering redundant information. As a result, the application of heterogeneous data fusion in TSE is getting popular. With the ability to record and share individual vehicle data, CAVs can serve as system-wide sensors—leading to a Lagrangian specification of the traffic flow field. In the near future, when the market penetration rate of CAVs is expected to be low, raw data from in-stream sensors (CAVs) can be fused with data from traditional sensors to achieve broader coverage and resiliency. Moreover, emerging technologies such as UAVs are expected to supply sufficient data, making heterogeneous data fusion more relevant in current circumstances. Heterogeneous data fusion techniques are widely adopted for estimating traffic parameters such as travel time and traffic density. In the following sections, we focus on state-of the- art methods for TSE, particularly in the context of heterogeneous data fusion in estimating travel time and traffic density.

8.11.1 TRAVEL TIME ESTIMATION

Travel time is a critical traffic parameter as it helps road users decide on route choices. Numerous studies have been carried out to exploit disparate heterogeneous data sets for travel time estimation. The estimation methods utilized for this objective can be classified into three categories: (1) statistical models, (2) probability-based models, and (3) artificial intelligence-based models [49]. Probe vehicle data are a common means to supplement loop detector data in travel time estimation. Much research has been performed to utilize loop detector data along with probe vehicle data. In Ref. [50], it is used GPS-equipped bus as probe vehicle along with loop detector and mobile phone network data to estimate travel time. The work assessed the contribution of each data source toward travel time estimation. The study concluded that data fusion does not necessarily improve estimation results all the time. However, in dense urban areas, the fusion of bus GPS and loop detector improved travel time estimation compared to the sole use of loop detector. Travel time estimation on urban road poses a significant challenge as the process is highly stochastic. Loop detector data and probe vehicle data on urban arterial are often inconsistent and sometimes even contradictory. An iterative Bayesian fusion method was applied in Ref. [51] to counter the challenge. Estimation was found to be more accurate than loop detector only or probe vehicle data only approaches. It was also more accurate than a single Bayesian fusion method-based travel time estimation.

8.11.2 TRAFFIC DENSITY ESTIMATION

Spatiotemporal distribution of traffic density helps in determining traffic congestion and design appropriate control algorithms. The traditional way of estimating traffic density is performed via loop detector or video processing. These approaches can estimate average traffic density only, as fixed sensors do not cover the entire road segment. Supplementing fixed sensors data with probe vehicle data is an effective measure to solve this problem. Loop detector and probe vehicle data were integrated in Ref. [52], to determine the density of traffic in freeway. Another study showed that loop detector and probe vehicle data fused with Rao-Blackwellized particle filter could improve traffic density estimation by 30% in comparison to the case where only loop detector data was used [53].

Different versions of the Kalman filter are commonly used as probabilistic state estimators. In Ref. [54], data form loop detector, automated vehicle identification (AVI), GPS location samples, and Bluetooth travel time readings were used for TSE on a homogeneous freeway segment. In Ref. [55], real-time TSE was performed by fusing loop detector data, GPS probe vehicle data, and Bluetooth scanner data. An extended Kalman filter was used in this work to synchronize time steps and sampling period, and heterogeneous data were fused with a Bayesian observer. In Ref. [56], a TSE model was proposed, which uses Lagrangian-space Kalman filter along with travel time transition model (TTM). This work formulates a state-space version of TTM using the framework of Kalman filter. The advantage of using the TTM over the traditional cell transmission model (CTM) is cost-effectiveness. Moreover, TTM-based estimators can recognize congestion better than CTM-based ones. Various machine learning-based methods have also been applied in TSE. However, these data-driven approaches are only accurate for a specific road segment. These approaches are not suitable for generalized application scenarios.

8.12 ITS CONTRIBUTION TO THE SOLUTION OF GLOBAL ISSUES

8.12.1 ITS AND ENVIRONMENT PROTECTION

ITS and public transport Local pollution. Despite success in arresting the negative trends of air pollution, the challenge remains huge, especially with regard to noise pollution. In Europe, for example, a quarter of the population lives less than 500 m from a road carrying more than 3 million vehicles per year. Consequently, nearly 4 million life-years are lost each year due to pollution [4]. Climate change mitigation. Although transport is not the primary global polluter, it is a considerable source of Green House Gases (GHG) and within this of CO_2 emissions. With the current rates of emissions, CO_2 concentrations will likely double their pre-industrial level by the end of the 21st century. Clearly, any transport policy considerations should address climate change. Furthermore, transport decision makers need to be able to measure traffic-induced GHG. ITS solutions can be instrumental in this regard, as well. For this to happen, a lead agency or cooperation among the key stakeholders is warranted. The Ministerial Conference on Global Environment and Energy in Transport (MEET), held in Tokyo (Japan) in January 2009, as well as MEET 2010, held in Rome (Italy) in November 2010, shared the long-term vision of the World Harmonization Forum of Vehicle Regulations (UNECE WP.29) in achieving low-carbon and low-pollution transport systems, which also ensure sustainable development. The ministerial declaration encouraged countries to broaden the diffusion and transfer of existing technologies and encourage research, development and the deployment of innovative technologies and measures such as ITS. More broadly, the draft decision of the Copenhagen Accord 2009, as well as the Cancun Agreement 2010 within the framework of the United Nations Climate Change Conference (UNFCCC), recommend various approaches to climate change, including opportunities to use markets, enhance the cost-effectiveness, and promote mitigation actions. Imagine that the transportation sector succeeds in renewing its technological base and managing its growth in a climate-neutral way, while meeting the mobility demand [57].

8.12.2 ITS AND PUBLIC TRANSPORT

Making public transport available, affordable and attractive is among the key transport policy goals. ITS, with its capacity to bring real-time information to travelers, can be an important player in achieving this goal.

8.12.3   ITS AND THE GLOBAL ROAD SAFETY CRISIS

Following the declaration of the First Global Ministerial Conference on Road Safety held in Moscow in November 2009, the United Nations General Assembly declared 2011–2020 as the "Decade of Action for Road Safety", with the goal to stabilize and then reduce the forecast level of global road deaths by 2020. Since the first motor vehicle was put into operation, around 30 million lives have been lost in road traffic accidents. Globally, 1.3 million people are killed on roads and 50 million more injured every year. Traffic accidents are often seen as personal and family tragedies, but in fact they are also tragic for society as a whole. Taking into account the direct economic costs of road crashes alone, the costs are estimated to be around US\$ 518 billion globally every year. At the same time, we should be realistic: with every day there are more people on the planet and they travel more. To address the global road traffic safety crisis, many more Governments are committed to take actions than ever before in history. They will—hopefully—take a system approach and implement the most appropriate policies and measures. To successfully combat the road safety crisis, it is imperative to put all resources to their maximum use, including the mainstreaming of ITS solutions [57].

REFERENCES

1. Guerrero-Ibáñez, J., Zeadally, S., & Contreras-Castillo, J. (2018). Sensor technologies for intelligent transportation systems. *Sensors*, 18(4), 1212.
2. Guerrero-Ibanez, J. A., Zeadally, S., & Contreras-Castillo, J. (2015). Integration challenges of intelligent transportation systems with connected vehicle, cloud computing, and internet of things technologies. *IEEE Wireless Communications*, 22(6), 122–128.
3. Mahmood, A., Siddiqui, S. A., Sheng, Q. Z., Zhang, W. E., Suzuki, H., & Ni, W. (June 2022). Trust on wheels: towards secure and resource efficient IoV networks. *Computing*, 104(6), 1337–1358. doi: 10.1007/s00607-021-01040-7.
4. Guerrero-Ibáñez, J. A., Zeadally, S., & Contreras-Castillo, J. (2015). Integration challenges of intelligent transportation systems with connected vehicle, cloud computing, and internet of thing technologies. *IEEE Wireless Communications*, 22, 122–128.
5. Bechler, M., Jaap, S., & Wolf, L. (2005). An optimized tcp for internet access of vehicular ad hoc networks. *Proceedings of the 4th International IFIP-TC6 Networking Conference (NETWORKING'05)*. doi: 10.1007/11422778_70
6. Bachir, A., & Benslimane, A. (2003). A multicast protocol in ad hoc networks inter-vehicle geocast. *Proceedings of the 57th IEEE Semiannual Vehicular Technology Conference. VTC 2003-Spring*, Jeju, South Korea.
7. Lee, U., & Gerla, M. (2010). A survey of urban vehicular sensing platforms. *Computer Networks*, 54, 527–544.
8. Dmitriev, I., & Kirillov, A. (2017). Smart roads and intellectual transport system. *Construction of Unique Buildings and Structures*, 2(53), 7–28. doi: 10.18720/CUBS.53.1.
9. Khamukhin, A. A., & Bertoldo, S. (2016). Spectral analysis of forest fire noise for early detection using wireless sensor networks. In *Proceedings of the International Siberian Conference on Control and Communications (SIBCON)* (pp. 1–4.). Moscow, Russia. 12–14 May 2016., pp. 1–4, doi: 10.1109/SIBCON.2016.7491654.
10. Bolourchi, P., & Uysal, S. (2013). Forest fire detection in wireless sensor network using fuzzy logic. In *Proceedings of the 2013 Fifth International Conference on Computational Intelligence, Communication Systems and Networks* (pp. 83–87). Madrid, Spain. 5–7 June 2013.. doi: 10.1109/CICSYN.2013.32.

11. Guerrero Ibáñez, J. A., Cosío-Leon, M., Espinoza Ruiz, A., Ruiz Ibarra, E., Sanchez López, J., Contreras-Castillo, J., & Nieto-Hipolito, J. (2017). GeoSoc: a geocast-based communication protocol for monitoring of marine environments. *IEEE Latin America Transactions*, 15, 324–332. doi: 10.1109/TLA.2017.7854629.

12. Albaladejo Pérez, C., Soto Valles, F., Torres Sánchez, R., Jiménez Buendía, M., López-Castejón, F., & Gilabert Cervera, J. (2017). Design and deployment of a wireless sensor network for the Mar Menor Coastal Observation System. *IEEE Journal of Oceanic Engineering*, 1–11. doi: 10.1109/JOE.2016.2639118.

13. USA Today NHTSA to Require Backup Cameras on All Vehicles (accessed on 11 October 2017). Available from: https://www.usatoday.com/story/money/cars/2014/03/31/nhtsa-rear-view-cameras/7114531/.

14. Automotive Sensors and Electronics Expo (accessed on 11 October 2017). Available from: https://www.st.com/content/st_com/en/about/events/events.html/automotive-sensors-electronics-expo.html.

15. Fleming W. J. (2008). New automotive sensors—a review. *IEEE Sensors Journal*, 8, 1900–1921. doi: 10.1109/JSEN.2008.2006452.

16. Abdelhamid, S., Hassanein, H. S., & Takahara, G. (2014). Vehicle as a mobile sensor. *Procedia Computer Science*, 34, 286–295. doi: 10.1016/j.procs.2014.07.025.

17. Contreras, S., Kachroo, P., & Agarwal, S. (2015). Observability and sensor placement problem on highway segments: a traffic dynamics-based approach. *IEEE Transactions on Intelligent Transportation Systems*, 17(3), 848–858.

18. Bargagli, B., Manes, G., Facchini, R., & Manes, A. (2012). Acoustic sensor network for vehicle traffic monitoring. *Proceedings of the First International Conference on Advances in Vehicular Systems, Technologies and Applications*, Venice, Italy. 24–29 June 2012.

19. Mathew, T. V. (2014). Transportation Systems Engineering (accessed on 11 October 2017). IIT Bombay. Available from: https://nptel.ac.in/downloads/105101008/.

20. Potter, T., & Reno, A. (2005). *The Evolution of Inductive Loop Detector Technology*. Reno A&E.

21. Wang, Y., & Nihan, N. L. (2003). Can single-loop detectors do the work of dual-loop detectors? *Journal of Transportation Engineering*, 129(2), 169–176.

22. Zhou, Y., Dey, K. C., Chowdhury, M., & Wang, K. C. (2017). Process for evaluating the data transfer performance of wireless traffic sensors for real-time intelligent transportation systems applications. *IEEE Transactions on Intelligent Transportation Systems*, 11, 18–27. doi: 10.1049/iet-its.2015.0250.

23. Ahmad, F., Basit, A., Ahmad, H., Mahmud, S. A., Khan, G. M., & Yousaf, F. Z. (2013). Feasibility of deploying wireless sensor based road side solutions for Intelligent Transportation Systems. In *Proceedings of the 2013 International Conference on Connected Vehicles and Expo (ICCVE)* (pp. 320–326). Las Vegas, NV. 2–6 December 2013.doi: 10.1109/ICCVE.2013.6799814.

24. Geetha, S., & Cicilia, D. (2017). IoT enabled intelligent bus transportation system. In *Proceedings of the 2017 2nd International Conference on Communication and Electronics Systems (ICCES)* (pp. 7–11). Coimbatore, India. 19–20 October 2017. doi: 10.1109/CESYS.2017.8321235.

25. Finogeev, A., Finogeev, A., Fionova, L., Lyapin, A., & Lychagin, K. A. (2019). Intelligent monitoring system for smart road environment. *Journal of Industrial Information Integration*, 15, 15–20.

26. Guerrero-Ibáñez, J., Zeadally, S., & Contreras-Castillo, J. (2018). Sensor technologies for intelligent transportation systems. *Sensors*, 18(4), 1212.

27. Contreras-Castillo, J., Zeadally, S., & Guerrero-Iban˜ez, J. A. (2017). Internet of vehicles: architecture, protocols, and security. *IEEE Internet of Things Journal*, 5(5), 3701–3709.

28. Abdelhamid, S., Hassanein, H. S., & Takahara, G. (2014). Vehicle as a mobile sensor. *Procedia Computer Science*, 34, 286–295.

29. Guerrero-Ibáñez, J., Zeadally, S., & Contreras-Castillo, J. (2018). Sensor technologies for intelligent transportation systems. *Sensors*, 18(4), 1212.

30. Guerrero-Ibáñez, J. A., Flores-Cort es, C., & Zeadally, S. (2013). Vehicular ad-hoc networks (VANETs): Architecture, protocols and applications. In *Next-Generation Wireless Technologies* (pp. 49–70). Springer. https://doi.org/10.1007/978-1-4471-5164-7_5.

31. Department of Transportation. (2007). National Highway Traffic Safety Administration. Technical Report: Federal Motor Vehicle Safety Standards. Tire Pressure Monitoring Systems; Controls and Displays (accessed on 14 March 2018). Available from: https://www.nhtsa.gov/sites/nhtsa.dot.gov/files/fmvss/TPMSfinalrule.pdf.

32. Ziraknejad, N., Lawrence, P. D., & Romilly, D. P. (2015). Vehicle occupant head position quantification using an array of capacitive proximity sensors. *IEEE Transactions on Vehicle Technology*, 64, 2274–2287. doi: 10.1109/TVT.2014.2344026.

33. Volvo Car City Safety w/Collision Warning (accessed on 27 March 2018). Available from: https://www.volvocars.com/en-om/support/car/xc40/article/d9330ec784320125c0a 801512e728d7b.

34. Dedicated Short Range Communications (DSRC) Home. Available from: leearmstrong. com. Archived from the original on 10 November 2012. Retrieved 29 February 2008.

35. Andert, E., Khayatian, M., & Shrivastava, A. (2017). Crossroads. In *Crossroads: Time-Sensitive Autonomous Intersection Management Technique* (pp. 1–6). Institute of Electrical and Electronics Engineers Inc. doi: 10.1145/3061639.3062221.

36. Khan, N., Mis˘i c, J., & Mis˘i c, V. B. (2016). VM2M: an overlay network to support vehicular traffic over LTE. In *2016 International Wireless Communications and Mobile Computing Conference (IWCMC)* (pp. 13–18). IEEE, Paphos, Cyprus. doi: 10.1109/ IWCMC.2016.7577026.

37. Reyes M uñoz, M. A., Barrado Muxı´, C., & Guerrero Ibañez, J. A. (2016). Communication technologies to design vehicle-to-vehicle and vehicle-to-infrastructures applications. *Latin American Applied Research*, 1(46), 29–35.

38. Kachroo, P., Agarwal, S., Piccoli, B., & Ozbay, K. (2017). Multiscale modeling and control architecture for V2X enabled traffic streams. *IEEE Transactions on Vehicular Technology*, 66(6), 4616–4626.

39. Silva, C. M., Masini, B. M., Ferrari, G., & Thibault, I. (2017). A survey on infrastructure-based vehicular networks. *Mobile Information Systems*, 2017, 6123868.

40. Cunha, F., Villas, L., Boukerche, A., Maia, G., Viana, A., Mini, R. A. F., & Loureiro, A. A. F. (2016). Data communication in VANETs: protocols, applications and challenges. *Ad Hoc Networks*, 44, 90–103.

41. Golestan, K., Soua, R., Karray, F., & Kamel, M. S. (2016). Situation awareness within the context of connected cars: a comprehensive review and recent trends. *Information Fusion*, 29, 68–83.

42. Grumert, E. F., & Tapani, A. (2018). Traffic state estimation using connected vehicles and station ary detectors. *Journal of Advanced Transportation*, 2018, 4106086.

43. Bonomi, F. (2013). The smart and connected vehicle and the Internet of Things. In *Invited Talk, Workshop on Synchronization in Telecommunication Systems*. https:// tf.nist.gov/seminars/WSTS/PDFs/1-0_Cisco_FBonomi_ConnectedVehicles.pdf

44. Miche, M., & Bohnert, T. M. (2009). The internet of vehicles or the second generation of telematic services. *ERCIM News*, 77, 43–45.

45. Agarwal, S., & Kachroo, P. (2019). Controllability and observability analysis for intelligent transportation systems. *Transportation in Developing Economies*, 5(1), 2.

46. Agarwal, S., Kachroo, P., & Contreras, S. (2015). A dynamic network modeling-based approach for traffic observability problem. *IEEE Transactions on Intelligent Transportation Systems*, 17(4), 1168–1178.

47. Seo, T., Bayen, A. M., Kusakabe, T., & Asakura, Y. (2017). Traffic state estimation on highway: a comprehensive survey. *Annual Reviews in Control*, 43, 128–151.
48. Contreras, S., Kachroo, P., & Agarwal, S. (2015). Observability and sensor placement problem on highway segments: a traffic dynamics-based approach. *IEEE Transactions on Intelligent Transportation Systems*, 17(3), 848–858.
49. Lim, S., & Lee, C. (2011). Data fusion algorithm improves travel time predictions. *IEEE Transactions on Intelligent Transportation Systems*, 5(4), 302–309.
50. Liu, K., Cui, M.-Y., Cao, P., & Wang, J.-B. (2016). Iterative Bayesian estimation of travel times on urban arterials: fusing loop detector and probe vehicle data. *PLoS One*, 11(6), e0158123.
51. Liu, N. (2011). Internet of vehicles: your next connection. *Huawei WinWin*, 11, 23–28.
52. Zhu, L., Guo, F., Polak, J. W., & Krishnan, R. (2017). Multi-Sensor Fusion Based on the Data from Bus GPS, Mobile Phone and Loop Detectors in Travel Time Estimation (Tech. Rep.).
53. Wright, M., & Horowitz, R. (2016). Fusing loop and GPS probe measurements to estimate freeway density. *IEEE Transactions on Intelligent Transportation Systems*, 17(12), 3577–3590.
54. Deng, W., Lei, H., & Zhou, X. (2013). Traffic state estimation and uncertainty quantification based on heterogeneous data sources: a three detector approach. *Transportation Research Part B: Methodological*, 57, 132–157.
55. Nantes, A., Ngoduy, D., Bhaskar, A., Miska, M., & Chung, E. (2016). Real-time traffic state estimation in urban corridors from heterogeneous data. *Transportation Research Part C: Emerging Technologies*, 66, 99–118
56. Yang, H., Jin, P. J., Ran, B., Yang, D., Duan, Z., & He, L. (2019). Freeway traffic state estimation: a Lagrangian-space Kalman filter approach. *Journal of Intelligent Transportation Systems*, 23(6), 525–540.
57. United Nations Economic Commission for Europe. *Intelligent Transport Systems (ITS) for Sustainable Mobility*. ASU Library, Tempe, AZ 85287.

9 The Future of Intelligent Transport Systems

9.1 BACKGROUND

It poses enormous challenges in terms of the efficient management of cities. One of the main components of the city infrastructure is its transportation system which often battles with problems such as road congestion, delays, crashes, and emissions. Also, Transportation systems have become a fundamental base for the economic growth of all nations. Nevertheless, many cities around the world are facing an uncontrolled growth in traffic volume, causing serious problems such as delays, traffic jams, higher fuel prices, increases in CO_2 emissions, accidents, emergencies, and the degradation of quality of life in modern society. According to a report by the Texas Transportation Institute, in the United States, commuters spend approximately 42 h a year stuck in traffic, drivers waste more than 3 billion gallons of fuel per year, having a total nationwide price tag of $160 billion, equivalent to $960 per commuter [1]. Such problems will worsen in the future because of population growth and the increasing migration to urban areas in many countries around the world as reported by the United Nations Population Fund [2] and the Population Reference Bureau [3]. Hence, there is a strong need to improve the safety and efficiency of transportation.

9.2 WHAT IS AN INTELLIGENT TRANSPORTATION SYSTEM?

An intelligent transportation system (ITS) is an advanced application that aims to provide innovative services relating to different modes of transport and traffic management and enable users to be better informed and make safer, more coordinated, and 'smarter' use of transport networks [4]. ITS also called smart transportation, is one such innovative concept that enables reliable and more personalized travel experiences to move around in cities. ITS can revolutionize the way people commute in metros and smart cities. ITS offers a novel approach to providing different transportation modes, advanced infrastructure, traffic, and mobility management solutions. It uses a number of electronics, wireless, and communication technologies to provide consumers access to a smarter, safer, and faster way to travel. ITS strives to innovate, plan, operate, evaluate, and manage transportation systems by leveraging advanced information and communication technologies. ITS refers to the use of technology to collect and analyze information related to the sector to deliver integrated transportation solutions. It focuses on various modes of transportation, infrastructure, vehicles, traffic management, stakeholders, and smart mobility. From a holistic point of view, it rectifies errors related to transportation, and infrastructure and enables systematic management of the entire transport system by leveraging a wide range of technology.

DOI: 10.1201/9781032691787-9

FIGURE 9.1 Intelligent transportation systems.

It is one of the important components of many innovative transportation solutions like mobility as a service, Connected, and automated mobility. Figure 9.1 shows ITS.

According to the US Department of Transportation, "Intelligent Transportation Systems (ITS) apply a variety of technologies to monitor, evaluate, and manage transportation systems to enhance efficiency and safety." Putting visions of science fiction-style transportation aside for the moment, this definition can be simplified into the following concepts for what makes up smart transportation: management, efficiency, and safety. In other words, smart transportation uses new and emerging technologies to make moving around a city more convenient, more cost-effective (for both the city and the individual), and safer. On the other hand, ITS is a combination of leading-edge information and communication technologies used in transportation and traffic management systems to improve the safety, efficiency, and sustainability of transportation networks, reduce traffic congestion, and enhance drivers' experiences. The possibilities are endless. ITS integrates advanced sensing, communication, and control technologies into transportation infrastructure and vehicles. In other words, ITS is an advanced application that aims to provide innovative services relating to different modes of transport and traffic management and enable users to be better informed and make safer, more coordinated, and 'smarter' use of transport networks [1]. Some of these technologies include calling for emergency services when an accident occurs and using cameras to enforce traffic laws or signs that mark speed limit changes depending on conditions. Although ITS may refer to all modes of transport, the directive of the European Union 2010/40/EU, made on July 7, 2010, defined ITS as systems in which information and communication technologies are applied in the field of road transport, including infrastructure, vehicles, and users, and traffic management and mobility management, as well as for interfaces with other modes of transport [4]. ITS may be used to improve the efficiency and safety of transport in a number of situations, i.e. road transport, traffic management, mobility, etc. [5]. ITS technology is being adopted across the world to increase the capacity of busy roads and reduce journey times [6]. ITS enables many applications to improve traffic flow, avoid accidents, and increase passengers' safety and comfort. ITS has the potential to reduce environmental impacts and improve quality of life significantly. However, they need to be widely deployed before reaching their full potential. ITS applications have four general categories: traffic control, safety, driver assistance, and infotainment. In the following sections, we explain each and provide examples. ITS deals with exploring emerging technologies and system engineering techniques to develop and enhance all types of transportation systems [7–9].

Moreover, effective use of infrastructure, capacity, and technology in ITS requires a lot of planning well in advance by ITS specialists. That can be implemented by collaboration or public-private partnerships. Because there are so many things that need to be taken into consideration while implementing ITS, e.g., transport modes, design, routing, vehicles, technology type, and traffic flows, to make transportation safe and well-coordinated.

9.3   WHAT IS THE NEED FOR ITS?

Transport authorities continue to raise the bar for safe and hassle-free transportation for commuters, but there are other challenges that commuters face related to urban congestion, inadequate road infrastructure, aging infrastructure, road safety, inefficient public transportation, and higher energy consumption. ITS can play an important role in solving these problems and better managing and controlling transportation systems in real-time. ITS facilitates new opportunities and more transportation choices integrated with easy-to-use technology. It is a multi-disciplinary concept that presents much-needed and cost-effective transportation solutions for smart cities. ITS can use resources and infrastructure effectively (existing as well as new), Plan, design, and implement comprehensive transportation systems, offer multi-modal, adequate, and on-demand transportation options, enhance public transportation management and its attractiveness, combat urban congestion, improve road safety and security, reduce fuel and energy consumption levels, Control and manage traffic in the cities, Make transport safe, efficient, manageable, and sustainable. The research and development on ITS have various implications, such as improving transportation convenience and safety (e.g., autonomous driving), enhancing efficiency in transportation (e.g., traffic congestion reduction), and having transportation schemes for environmentally-friendly traffic (e.g. by minimizing fuel consumption) since we are very much dependent on transportation systems in our day-to-day lives [4]. ITS focus on integrating various technologies with vehicles and transportation infrastructure to make transportation safer, cheaper, and more efficient by applying various technologies. ITS could provide five key merits by (1) enhancing traffic safety, (2) improving transportation network performance, especially through traffic jam reduction, (3) increasing traffic flow and comfort, (4) providing green environments, and (5) raising the production and lifting the growth in employment as well as economy [10, 11].

ITS mainly takes advantage of state-of-the-art technologies such as wireless and satellite communication, mobile technology, surveillance systems, electronic sensors and detectors, digital image processing software, geo-positioning navigation systems, etc. to better manage the transportation system.

9.4 APPLICATIONS INSIGHTS OF ITS

The global ITS market was valued at USD 26.58 billion in 2019 and is expected to grow at a CAGR of 5.8% from 2020 to 2027. Various organizations and countries like the USA, Germany, and Japan have increased their investments in ITS concerning safe and smart mobility factors. ITS mainly takes advantage of state-of-the-art technologies such as wireless and satellite communication, mobile technology, surveillance systems, electronic sensors and detectors, digital image processing software, geo-positioning navigation systems, etc. to better manage the transportation system. ITS can be deployed in several applications; some of which are mentioned below:

1. Traffic Management and Operations: Get real-time information related to traffic. It predicts traffic conditions and also responds to changing traffic conditions.
2. Vehicle Control and Speed Limit: Improve driver control by alerting about possible imminent vehicle collisions and the speed limit on the highway.
3. Traveler Information: Get all the data of transportation in a system to assist travelers with route planning, ETAs, emergencies, route guidance, etc.
4. Ramp Meter: Detect traffic density and regulate the flow of vehicles entering on a freeway on-ramp and mainline both.
5. Commercial Vehicle operation: Monitor the operations of commercial vehicles like trucks and provide safety to en-route travelers.
6. Public Transportation Management: Get real-time passenger as well as vehicle information, arrival time, location, approaching vehicle, etc., for efficient operations.
7. Electronic Toll Collection: Accelerate the payment process on toll stations with automatic vehicle identification, classification, and positioning system to alleviate inefficiencies.
8. Variable Message Signs: Update travelers via digital sign boards about traffic congestion, collisions, speed limit, or any event. Other applications include parking management, traffic signal coordination, transit signal priority, traffic cameras, weather monitoring, traffic data collection, traffic data centers, freight and fleet management, vulnerable user systems, and multi-modal trip planning.

9.5 HOW ITSs WORK?

The state-of-the-art technologies have a major influence on the functioning of ITS. Let us understand how it works:

Data acquisition: Automatic data acquisition using high-level hardware devices and software is the first stage here for strategic planning and implementation of ITS.

Transport specialist collects data related to traffic, infrastructure, and vehicles. Technologies used at this stage are sensors, detectors, video image processors, Global Positioning System (GPS), vehicle identifiers, etc. to obtain traffic flows, congestion, and vehicle-related situations and to make suitable decisions further.

Data processing: At this stage, data specialists like data scientists or engineers verify, reconcile, and consolidate the data in a single place for further meaningful interpretation. By using modern data processing software and devices like GPS, sensors, and detectors, real-time data processing occurs with the highest accuracy of the outcome.

Data analysis: All the data undergoes a procedure from data cleaning to data fusion and logical analysis with the help of software and hardware for more accurate transport system planning and to predict traffic scenarios.

Data distribution: The system delivers real-time transport information like travel time, speed, delay, traffic-related parameters, vehicle operation, etc., through various ways, broadcasting or electronic devices, and mobile-based technologies in order to improve efficiency and safety.

9.6   MAIN BENEFITS OF ITS

ITS contributes to making transportation easier and cities smarter. It enhances the travel experience by enabling real-time transit data, traffic flows, emergencies, etc. with the help of advanced communication technologies. It offers several solutions for better transport management. ITS benefits fall into mainly three categories as described below:

1. Safety: Powered with the latest technology like IoT, computer vision, and machine learning, ITS optimizes transport-related data and operational processes. Furthermore, these technologies have the potential to integrate users, infrastructure, and vehicles on a single platform that enables the transport authorities to monitor real-time traffic scenarios, incidents, or collisions and respond promptly. ITS alerts the drivers on their roadways so that they can limit their speed or change route, which reduces congestion and preempts incidents, and enhances public safety.
2. Efficiency: ITS is all about data collection, analysis, and interpretation to make the transport system efficient and manageable. Automated technologies let the authorities control and monitor the road network, traffic flows, vehicle speed, parking opportunities, etc. consistently. This helps to identify real-time problems or any transport-related needs, which authorities can fix as soon as possible and continue managing the transport system effectively.
3. Sustainability: ITS holds the promise of enabling sustainable transportation for smart cities. It encourages public transportation usability and punctuality. The higher the use of public transport, the lesser the greenhouse gas emissions in the environment. Advanced traffic incident management system in ITS quickly responds to emergencies, which reduces travel times and congestion, and ultimately air-borne pollutants, promoting sustainable transportation systems.

Other Benefits include the following: new business opportunities because of the adoption of collaborative and public-private partnerships that result in higher employment opportunities. Improves the transportation of goods by controlling and managing the movement of commercial vehicles. Fosters innovations in the field of transportation through the development and implementation of advanced technologies and innovative field concepts. Creates economic value by seamlessly interconnecting and utilizing various components of the sector e.g. infrastructure. Meets all the accessibility and transportation-related needs of those making use of transport networks in smart cities. Many countries are taking advantage of the intelligent transport system to solve mobility issues. The ultimate benefit of implementing this concept is to manage traffic flows, prevent potential incidents, and decrease the negative impact on the environment. State-of-the-art technologies improve the overall performance of the transportation system. Transport authorities can take advantage of ITS for current transportation management and future planning of the transportation systems of the cities. Smart and intelligent ITS promise to address issues such as high fuel prices, high levels of CO_2 emissions, high levels of traffic congestion, and improved roads [12, 13].

9.7 INTELLIGENT TRANSPORT SYSTEMS TECHNOLOGIES

Intelligent transport systems vary in technologies applied, from basic management systems such as car navigation; traffic signal control systems; container management systems; variable message signs; automatic number plate recognition, or speed cameras to monitor applications, such as security CCTV systems, and automatic incident detection or stopped vehicle detection systems; to more advanced applications that integrate live data and feedback from a number of other sources, such as parking guidance and information systems; weather information; bridge de-icing (US deicing) systems; and the like. Additionally, predictive techniques are being developed to allow advanced modeling and comparison with historical baseline data. Some of these technologies are described in the following sections [14].

Various forms of wireless communications technologies have been proposed for ITSs. Radio modem communication on UHF and VHF frequencies is widely used for short- and long-range communication within ITS. Short-range communications of 350 m can be accomplished using IEEE 802.11 protocols, specifically 802.11p (WAVE) or the dedicated short-range communications (DSRC) 802.11bd standard being promoted by the Intelligent Transportation Society of America and the United States Department of Transportation. Theoretically, the range of these protocols can be extended using mobile ad hoc networks or mesh networking. Longer-range communications use infrastructure networks such as 5G. Long-range communications using these methods are well established, but, unlike the short-range protocols, these methods require extensive and very expensive infrastructure deployment.

Recent advances in-vehicle electronics have led to a move toward fewer, more capable computer processors on a vehicle. A typical vehicle in the early 2000s would have between 20 and 100 individual networked microcontroller/programmable logic controller modules with non-real-time operating systems. The current trend is toward fewer, more costly microprocessor modules with hardware memory management and real-time operating systems. The new embedded system platforms allow for more sophisticated software applications to be implemented, including model-based process control, artificial intelligence, and ubiquitous computing. Perhaps the most important of these for ITSs is artificial intelligence.

- Activities that were traditionally undertaken through human intervention can be automated.
- Road network performance can be monitored and adjusted, in real-time.
- Data that was previously collected by costly physical infrastructure can be provided through new, richer data sources.
- Analysis that was undertaken from historic data can now be undertaken by systems delivering intelligence through real-time data analytics.
- Road users' choices, previously influenced only through road signs, can be influenced through a wide array of publication channels such as mobile devices/in-car systems.
- These possibilities will only happen if Road Administrations have a clear and considered strategy for bringing existing disparate systems, services, and operational approaches together over time. ITS involves various emerging technologies and applications as shown in Figure 9.2.
- As illustrated in Figure 9.1, ITS can be categorized into two major areas: ITS Technologies and ITS Applications. ITS Technologies can be primarily grouped into Sensing Technology, Communication Technology, Computational Technology, and Actor-Network Technology. On the contrary, ITS applications can be grouped into five primary categories: Traveler Information Systems (TIS), Transportation Management Systems (TMS), Transportation Pricing Systems, Public Transportation Systems (PTS), and V2V/V2I-based ITS [14].

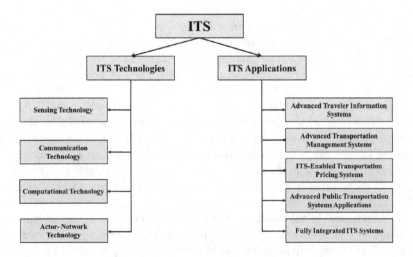

FIGURE 9.2 ITS involves technologies and applications.

9.7.1 Communication Technologies

Some of the key communication technologies used in ITS are as follows:

1. DSRC: DSRC is a short-range to medium-range wireless communication channel, operating in the 5.8 or 5.9 GHz spectrum, specially designed for automotive uses and is suitable for safety applications (such as post-crash notification) due to its low latency [15].
2. Wireless sensor networks: Wireless sensor networks, such as WiMAX [16], is a communication technology commonly used for wireless Internet access and allow rapid vehicular communications.
3. Mobile telephony: Although mobile telephony [17] is popular specially for urban areas and major roads, it may not be suitable for some safety-critical ITS applications, since it could be too slow.
4. Radio wave: Radio waves or infrared beacons are used in Japan to transfer real-time traffic information on arterial roadways, a moderate capacity roadway just below highways in the level of service.
5. Probe-vehicle: In some countries, taxis/government-owned vehicles equipped with DSRC or other wireless technology are often used as probe vehicles to send the probe data such as their locations and speed, to a centrally operated traffic management center (TMC). The TMC then aggregates the received data to get an impression of the traffic flow and congested locations of an area.

9.7.2 Computational Technologies

The benefits of this technology to measure traffic data are that it is less expensive and provides more coverage including nearly all locations and roadways, requires less maintenance, and is workable in all weather conditions. The three methods which have been used to collect the probe data are as follows [18].

1. Triangulation method: In the mid-2000s, in-vehicle mobile phones have tried to use anonymous traffic probes to receive instant traffic data as the vehicle moves. The traffic data can be transformed into traffic flow measurement by measuring and analyzing traffic data using triangulation and pattern matching or cell-sector statistics. The more congestion the more probes due to more cars and more phones. Ideally, the method should work better in case the distance between the antennas is shorter, such as in urban areas. The advantage of this method is that it does not require roadside units other than using only mobile phones. However, since the early 2010s, the usage of the triangulation method has been decreasing.
2. Vehicle re-identification: This method requires a number of detectors placed along the road. In this approach, a unique serial number of a device (e.g., MAC (Machine Access Control) addresses from Bluetooth devices or RFID (Radio-frequency identification) serial numbers/tags from Electronic Toll Collection (ETC) transponders) in the vehicle is detected once in one

location followed by detected again (reidentified) in another location of the road. Travel time and speed can be thereby estimated by measuring the time at which the specific device is detected by a pair of sensors.

3. GPS-based method: GPS is an increasingly popular method. Vehicles as equipped with in-vehicle GPS (satellite navigation) systems having two-way communication links to the traffic data provider. Here position data of the vehicles are obtained which are used to compute their speeds.

9.7.3 ACTOR-NETWORK TECHNOLOGY

It is evident that recently the interest of actors in ITS has paid a large attention. It is found that actor-network technology can play an important role in ITS to improve vehicle safety. An insecure, distributed, wireless vehicular safety system could be easily accessible resulting in poorer vehicular safety. Therefore, VANET safety messages need to be secured to mitigate falsified reports received from malicious vehicles and to detect erroneous information generated by untrusted vehicles. Precisely, safety messages should be authenticated to a vehicle, and revoke the credentials (i.e., the certificate(s)) of a malicious/untrusted vehicle (i.e., one that generates the falsified or erroneous messages). Three types of actors are considered for safety applications providing security and privacy in VANETs as described below.

1. Vehicles or On-Board Units (OBUs): Vehicles can be treated as intelligent actors. In safety applications, each vehicle can be assumed to be equipped with a temper-proof security device responsible for ensuring the privacy of the sensitive and personal information of the ego vehicle, such as private key, and to execute all the cryptographic operations that the vehicles need to operate, in order to participate to the secure vehicular network.

2. Certification Authorities (CAs): These entities represent the trust establishments. They are responsible for providing for each vehicle and RSU's personal certificate allowing it to prove its identity when communicating with other participants. A CA is also in charge of revoking certificates of untrusted and malicious nodes. It is responsible for establishing the security requirements within its region. Moreover, CAs cooperate to ensure interregions security (when vehicles move between regions managed by different CAs).

3. Roadside Units (RSUs): The RSU entities participate in the security architecture of the VANET; they are delegated by the corresponding CA to carry out some security functionalities, such as the generation of pseudo-certificates based on pseudonym pair of keys (i.e., public and private keys).

9.8 COOPERATIVE SYSTEMS ON THE ROAD

Communication cooperation on the road includes car-to-car, car-to-infrastructure, and vice versa. Data available from vehicles are acquired and transmitted to a server for central fusion and processing. These data can be used to detect events such as rain (wiper activity) and congestion (frequent braking activities). The server processes a driving recommendation dedicated to a single or a specific group of drivers and transmits it wirelessly to vehicles. The goal of cooperative systems is to use and plan communication and sensor infrastructure to increase road safety. The definition of cooperative systems in road traffic is according to the European Commission [9, 19]. "Road operators, infrastructure, vehicles, their drivers, and other road users will cooperate to deliver the most efficient, safe, secure, and comfortable journey. The vehicle-vehicle and vehicle-infrastructure cooperative systems will contribute to these objectives beyond the improvements achievable with stand-alone systems."

9.8.1 Smart Transportation - New Business Models

New mobility and smart transportation models are emerging globally. Bike sharing, car sharing and scooter sharing schemes like Lime or Bird are continuing to gain popularity; electric vehicle charging schemes are taking off in many cities; the connected car is a growing market segment; while new, smart parking solutions are being used by commuters and shoppers all over the world [20]. All these new models provide opportunities for solving last-mile issues in urban areas.

9.8.2 Traffic Monitoring and Control

Traffic monitoring and control constitute the central part of ITS application. It includes adapting traffic lights to optimize traffic flows, providing alternative or adaptive routes according to real-time traffic situations, coordinated ramp metering, automatic accident detection, and clearance, etc. These applications can make transportation efficient for both passengers and goods. For these applications to work correctly, it is essential to have a robust sensing mechanism consisting of synergy between multiple sources of information such as road sensors, vehicle sensors, and communication between vehicles and the road infrastructure [20].

9.8.3 Traffic Safety

Safety applications of ITS technologies include traffic flow management, automated speed management, video incident detection, and driver assistance systems, including adaptive cruise control, departure warning, and emergency brake assist (EBA). Another common element for managing traffic flow is using dynamic message signs. The drivers can get information from the road if there is a problem with the pavement, a traffic jam due to an accident, or simply slowing traffic ahead. This method is effective for simplifying the communication between road infrastructure, vehicles, and drivers. Advisory notices can also be sent to drivers through haptic systems (such as smart rear-view mirrors or front windshields). An example is automatic radars checking car speed at various locations and sending an advisory if the speed is over the limit. In addition, road cameras can take photographs of the vehicle at specific locations to recognize the car and send messages using road infrastructure and sensors within the vehicle [20].

9.8.4 Driver Assistance and Infotainment

In this application, the focus is on assisting drivers with a better driving experience. Applications include providing context information, remote diagnostics, navigation information, and real-time traffic alerts. Many sensors are deployed within the vehicle (lidar, radar, cameras) to sense the environment. However, the data from the individual sensor is often not significant; hence, we use many sources and combine them (sensor fusion). This combined information can create a synergy to provide the driver with a richer environmental perception. The applications include early detection of moving objects that could block the vehicle or conflict with its trajectory. Additionally, during congestion, driving assistance systems can help with repetitive tasks such as continuously breaking to keep a fixed headway with other vehicles or alert the driver if they detect tiredness or drowsiness in the driver. Infotainment applications can use intervehicle communication to share services among vehicles or roadside infrastructure. These applications bring intelligent services such as high-definition displays, touch controls, and powerful computers inside the car to provide drivers and passengers with multimedia driver information, video services, or even Internet connection [20].

9.8.5 Autonomous Driving

Due to excessive traffic, pollution, and limited fossil fuel resources; ride-sharing and vehicle path optimization have become the mainstream solution to these problems. This also has other major benefits, like the reduction of accidents due to human error, hence saving human lives. Millions of hours spent on the road can be saved by autonomous vehicles. Autonomous prototypes have also been found in other major use cases: production industry, warehouse management, vehicle platooning, waste collection, etc. The switch from manually driven to completely autonomous will not happen directly, rather will happen in different stages of autonomy. Advanced driver assistance systems (ADAS) have existed for more than a decade helping the driver in various tasks. Modern-day ADAS can perform a variety of functions, like self-parking, cruise control, traction control, antilock braking system, hill assist, sleep/drowsiness detection, reverse parking assist, etc. There are six levels of driving automation as explained in Table 9.1.

9.9 BENEFITS OF ITS

Minimize pollution: the aim of ITS is to promote the use of public transport in general masses. By providing single-point services and giving access to real-time information about transport schedules, delays consumers will be drawn toward public transport reducing private car usage thereby lowering traffic congestion and lowering pollution levels. Also, people will be motivated toward the use of clean fuel, bike sharing, and carpooling habits.

Security and safety: the real-time data analysis through GPS, CCTV, wireless and Internet connectivity, and advanced sensing technologies can help provide emergency and critical care services to drivers and travelers when required. Surveillance

TABLE 9.1

Autonomous Driving: Levels of Automation

Level	Automation	System
Level 0	No automation	The driver performs all primary operating tasks like steering, braking, accelerating or slowing down, and so forth
Level 1	Driver assistance	The vehicle can assist with some secondary and tertiary functions, but the driver still handles all primary tasks and monitoring of the surrounding environment
Level 2	Partial automation	The vehicle can assist with steering or acceleration functions and allow the driver to disengage from some of their tasks. The driver must always be ready to take control of the vehicle and is still responsible for most safety-critical functions and monitoring of the environment
Level 3	Conditional automation	The vehicle itself controls all monitoring of the environment (using sensors like LIDAR). The driver's attention is still critical at this level but can disengage from "safety-critical" functions like braking and leave it to the technology when conditions are safe
Level 4	High automation	The vehicle is capable of steering, braking, accelerating, monitoring the vehicle and roadway as well as responding to events, determining when to change lanes, turn, and use signals
Level 5	Complete automation	This level of autonomous driving requires absolutely no human attention. There is no need for pedals brakes, or a steering wheel, as the autonomous vehicle system controls all critical tasks, monitoring the environment and identification of unique driving conditions like traffic jams

of public transportation can help city managers in alerting against terror elements and avoid mishaps or terror attacks.

Smart parking solutions: parking woes affect every city dweller. Smart parking solutions with the help of the right infrastructure, Internet connectivity, security cameras can minimize them to a great deal. Many urban centers now have multi-layer parking system. Also, there are apps which guide users about the free parking space available nearby. While developed nations like United States, Europe, and Dubai have been investing heavily in ITS network for over a decade now. It is still a challenge for developing nations. The major problems faced by these countries in adoption of ITS network is lack of funding, lack of IT infrastructure, lack of formal transportation system, unplanned cities, illiteracy, poor public infrastructure etc. To emulate west, a lot has to be done in this arena from huge infrastructure funding to change in habits of general population and government's initiative for better transportation infrastructure for present and future generation.

Efficient traffic congestion management: ITS using a Machine learning framework can help access the real-time analytics and manage traffic congestion by getting prior information about traffic on a particular route, pre-emption alerts, and by accessing alternate routes and diversions. Additionally, congestion pricing, often known as congestion charges; is a method of taxing consumers of public goods during peak hours. It is a dynamic pricing approach, which is based on economic theory, is a frequent stratagem in the transportation sector to reduce traffic and pollution by charging more for accessing particularly crowded zones in major metropolitan cities. Based on the real-time data, such strategies will be helpful in improving the reliability of highway system performance and people's quality of life.

Efficient use of multiple top hatting: With the transport and logistics sector evolving hugely in the last decade, there is an emerging need to use vehicles without causing traffic congestion or excessive vehicular emissions. One or more vehicle upper body structures with the same platform are referred to as the top hat. Top hatting gives fleet owners an upper edge to utilize the same vehicle for different purposes at different time of the day. For instance, a delivery van at night can be utilized as a passenger bus during the day for ease of commuting. The analytics can help better understand the appropriate requirement of a vehicle at a given point of time, in a specific location and in the required quantity. This will lead to effective utilization of existing vehicles solving multiple purposes, eliminating the need to create large quantity of vehicles serving a single use case.

Driver management system: Another advantage of ITS is hassle-free driver assistance and management. The real-time data collected allows the fleet operators to get a complete understanding of their drivers, their productivity, and the general safety of the vehicle and driver through effective monitoring of the driver activities, including speeding, abrupt braking, and harsh braking. Also tracking the driving patterns, such as consecutive breaks and continuous driving, can help monitor drivers' fatigue levels so fleet owners can effectively arrange backup drivers if the current driver is incapacitated, as an immediate outcome. With consolidated driver patterns of tons of vehicles, the operator can use this data to plan driver shifts, distance and halts required for timely delivery services, in a much more efficient manner.

Smart fleet management: Deploying ITS for fleet management can help increase overall operational efficiency along with reducing congestion, effective multi-utility of vehicles, and driver fatigue levels. With the DNA of the vehicle at fingertips, fleet operators can set safety standards, prevent battery damage and overheating of the breaks, monitor material overloading and structural analysis, directly streamlining vehicle inspections. At EVage, we are reimagining mobility to transform lives and businesses by bringing innovation and tech prowess to the fore with a strong focus on elevating logistical experience.

Traveler information systems: TIS are used to guide the driver by sending various real-time information related to travel and traffic, such as bus/train routes and their time schedules as well as information regarding any delays caused by traffic jams/accidents/bad weather/road construction. This category also includes in-car navigation systems and telematics-based services, which offer a range of safety, route navigation, accidents notification, and concierge services, e.g. services regarding location, mobile calling, or in-vehicle entertainment with various options including access of Internet, downloads of music or movie, etc.

9.10   TRANSPORTATION MANAGEMENT SYSTEMS

Mainly provide ITS applications that concerns with traffic control devices for traffic signals, ramp metering, and dynamic message signs on highways for the purpose of real-time messaging about traffic or the status of the highway. It depends on centralized traffic management centers, called as Traffic Operations Centers (TOCs) run by cities and countries worldwide, to connect sensors and roadside equipment, probe vehicles, message signs, and other devices together to get an integrated overview of traffic flow and to detect vehicle crashes or any unexpected events occurred due to bad weather, or hazards on the road [20].

9.10.1 Transportation Pricing Systems

Transportation pricing systems play a major role in developing countries' ITS. They have been commonly used in ETC, where the drivers could pay tolls automatically through DSRC-enabled on-board device or ID placed on the windshield. Also, such systems can be used in High-Occupancy Toll (HOT) lanes reserved for buses and other high-occupancy vehicles where they only allow single occupant vehicles upon tool payment in case of traffic jam. In these reserved lanes, the transportation pricing systems could help to control the traffic flow (i.e. the number of vehicles) by deploying variable pricing (through ETC) to maintain a smooth traffic flow, even during rush hours [20].

9.10.2 Public Transportation Systems

PTS provide the automatic vehicle location (AVL) application for transit vehicles, i.e. bus or rail, to provide information about their current locations which make it possible for the traffic operations managers to get a real-time impression about the status of transit vehicles. PTS further make transit system more attractive by displaying the arrival and departure times of buses and trains providing particularly useful 'next bus' or 'next train' information to the passengers. PTS is also concerned with the electronic fare payment for public transport. V2V/V2I-based ITS systems combine both V2V (Vehicle-to-Vehicle) and V2I (Vehicle-to-Infrastructure) integration into a consolidated platform, would enable a number of driving assistance safety and non-safety ITS applications, including dynamic lane departure warnings, curve speed warnings, re-routing of traffic through variable message signs, or notify weather-related conditions, such as icing.

9.10.3 Major ITS Areas

The major areas of an ITS are the Intelligent Vehicle Initiative (IVI), the Commercial Vehicle Operations (CVO), and the Advanced Rural Transportation Systems (ARTS). The Intelligent Vehicle Initiative integrates driver assistance and information so that all vehicles will operate more safely. The CVO area has applications for commercial vehicles to speed the process of freight movement, carrier operations and vehicle inspections. The ARTS area will improve mobility, safety, efficiency, and communication in rural areas.

9.10.3.1 Intelligent Vehicle Initiative

The Intelligent Vehicle Initiative uses on-board collision avoidance systems based on radar or sonar technology. These systems detect objects that the vehicle may strike and warn the driver of the impending hazard. Dashboard monitors display travel maps and provide directions. An intelligent cruise control adapts vehicle speed to maintain a safe driving distance from other vehicles. During bad weather, onboard sensors warn drivers of obstacles in front of the vehicle and emit warning signals to help drivers avoid hitting objects. During slippery weather conditions, advanced vehicle and brake systems measure wheel speeds and steering wheel angles to determine if a spin is imminent. If it is, braking is automatically applied to the appropriate wheels, and the vehicle can balance out of the spin.

9.10.3.2 Commercial Vehicle Operations

The CVO area links carrier, state, and national information networks to facilitate a simple and cost-effective exchange of safety and business data. Commercial processes such as roadside safety inspections, credential checks, vehicle registration, fuel-tax collection, and hazardous materials transport are simplified. Electronic screening automates weight, safety and credential screenings at roadside weigh stations. International border clearances are also speeded along by this system. More unsafe drivers and vehicles can be removed from the road by this process, than by conventional methods. ITS technologies can also identify truckers with poor safety records for more frequent inspections, while compliant trucks are allowed to proceed.

9.10.3.3 Advanced Rural Transportation Systems

The ARTS applications help solve the transportation problems of the rural traveler. Speeders are common in rural areas, and advanced warned about animals crossing the road ahead. But when accidents do occur, response times are improved, even in remote, deserted areas thanks to a "panic button" in the vehicle. This system pinpoints the exact location of the accident and relays that information to local emergency agencies. Tourists and other travelers in the rural area may not be familiar with the roads, and directions may be hard to come by. In-vehicle navigation systems can get people back on the right road quickly. In rural areas, bad weather conditions and rugged terrain can combine to produce awful driving conditions. On-board weather updates can help save lives and property.

9.11   INTEGRATED COMPONENTS OF AN ITS

The main components of an ITS are Travel and traffic management, public transportation operations, electronic payment, CVO, advanced vehicle control and safety systems, emergency management, Information management, maintenance, and construction management. Also, there are nine integrated components of an ITS. These components include traffic signal control, freeway management, transit management, electronic fare payment, electronic toll payment, incident management, traveler information services, emergency management services and maintenance and construction management [21].

9.11.1   Traffic Signal Control Systems

Traffic signal control systems use in-pavement detectors to monitor the current demand. These systems measure the demand for right-of-way, shifts in directional demand, and changes in cross-street directional demand. The detectors relay this information to the traffic signals, which can then adapt to the current needs of the vehicles. The result is a smoother flow of traffic, with shorter waiting times. As a safety measure, automatic cameras at signalized intersections have reduced the number of speeders and red-light runners, by photographing license plate numbers and fining the violators.

9.11.2 Freeway Management Systems

Freeway Management Systems use ramp metering techniques to measure and regulate how much traffic is entering and leaving major freeways. The metering rate ensures that demand remains below capacity, reducing congestion. Metering rates also improve safety by breaking up groups of merging vehicles competing for space in the stream of traffic. Cities using Freeway Management Systems report handling more traffic while maintaining or increasing travel speeds.

9.11.3 Transit Management

Transit Management includes AVL technologies and computer-aided dispatch systems to help keep buses on schedule and improve service. Some cities have integrated the bus system with the traffic light system at key intersections. This increases the green lights along these routes by only a few seconds, but the end result is a reduction in transit travel time [22].

9.11.4 Electronic Fare Payment Technologies

Electronic Fare Payment Technologies have reduced cash-handling costs for transit operators, due to more accurate data collection. This method is also more convenient for passengers, who no longer have to worry about having the correct change for fare payment.

9.11.4.1 Electronic Toll Payment

Electronic toll payment allows travelers to speed through toll collection plazas without stopping to pay their toll. A roadside sensor locates a transponder in the approaching vehicle and automatically bills the corresponding account. This results in significantly reduced levels of vehicle emissions.

9.11.4.2 Incident Management Systems

Incident management systems include dynamic message signs, which alert travelers to accidents or stalled vehicles on the road ahead. Video cameras and road sensors help to detect and locate incidents quickly, and computer-aided dispatch can speed emergency services to the scene.

9.11.5 Traveler Information Services

Traveler Information Services utilize many modes of communication, such as the Internet, radio, kiosks, pagers, dial-up services, television, and on-board computers.

These services allow travelers to access pre-trip and en-route information so they can plan the most efficient route for their needs. Also available is information on all modes of public transportation and ride-sharing in the area, as well as specialized services for senior citizens or persons with disabilities [22].

9.11.6 EMERGENCY MANAGEMENT SERVICES

Emergency management services are greatly enhanced by traffic control centers that continually monitor roadway conditions. When an incident occurs, the nearest emergency service vehicle is located electronically and dispatched to the scene. Highway managers then alert other drivers of the incident through dynamic message signs. These services reduce response times, help save lives, and reduce the occurrence of secondary incidents [22].

9.12 TAXONOMY OF ITS APPLICATIONS

A taxonomy for ITS applications taxonomy defines six categories based on the type of application for ITS [23].

9.12.1 SAFETY CATEGORY

In the safety category, applications focus on improving the safety of drivers and passengers (as shown in Figure 9.3) thus reducing the number of accidents, injuries, and fatalities. An application for lane management focuses on keeping the vehicle safe while driving. Using cameras mounted behind the rear-view mirror the application can monitor road lane markings and detect any drifts outside of a lane [24]. Adaptive cruise control applications use radar, speed, and distance sensors to regulate the speed and maintain a safe distance from the vehicles in front. Sensors are combined with predictive algorithms such as neuro-fuzzy [25] or curve radius prediction [26] to determine the best angle for the turn in road curves. Blind spot information is an application that uses radar sensors to alert the driver when changing lanes if a car is detected in the blind spot zone [27]. Intersection collision warning applications

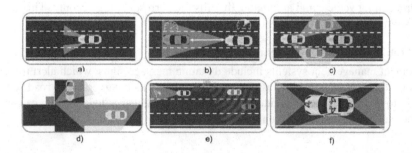

FIGURE 9.3 An example of ITS safety applications: (a) lane keeping aid, (b) adaptive cruise control, (c) blind spot information, (d) intersection collision warning, (e) road hazard warning, and (f) surround-view monitoring.

use position and speed sensors to determine the probability of vehicles colliding and transmits a warning signal when the probability of collision is higher than some established security range. Surround-view monitoring applications use cameras to detect obstacles around the car, making the parking and maneuvering process easier. To further improve safety applications, it is important to use data fusion from several sources and the use of intelligent processing algorithms to allow not only drivers' notifications but also to determine reaction times to make fast automated decisions and reduce the potential of road accidents.

9.12.2 Traffic Management Category

ITS applications in this category improve the traffic flow in roads and urban zones (as shown in Figure 9.4). Surveillance applications can be divided into two categories: fixed surveillance systems which consist of fixed stations that use cameras and sensors that are installed on the roads to monitor road conditions. The second category, called surveillance-on-the-road, uses sensors and cameras embedded in vehicles to support surveillance [28]. Lane management applications focus on managing the available capacity of the roads during special traffic conditions such as emergency evacuations, incidents, or risky weather [29]. This application uses RADAR, cameras, and infrared sensors to detect the occupancy, direction, and velocity of vehicles. Special event TMS are a variation of lane management systems and they are used to control and reduce road congestion problems at special places such as stadiums or convention centers. The usage of sensors (such as radar and infrared) and cameras allows a flow direction change in lanes based on traffic demands. Intersection management applications are cooperative applications that are a viable replacement for traditional traffic light-based methods for intersection control. In this application, road users, infrastructure and traffic control centers work in an integrated way,

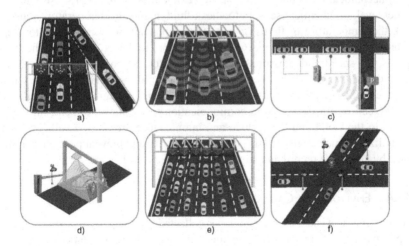

FIGURE 9.4 ITS traffic management applications, (a) lane management, (b) surveillance, (c) parking management, (d) automatic tolling, (e) special event transportation, and (f) intersection management.

combining RFID technology, proximity, ultrasonic, radar sensors, cameras, trajectory planning [30], and virtual traffic lights' modeling, to coordinate traffic safety more efficiently [31]. Parking Management Applications (PMA) use magnetometers, microwaves, inductive loops, infrared or RFID technologies to collect information about parking occupation and inform drivers about parking opportunities or available spaces near their zone [32], assisting in the management and distribution of parking spaces thereby reducing traveler frustration and congestion problems associated with searching for parking.

Traffic management applications are becoming increasingly important. However, it is imperative for all those applications to form an integral and collaborative system to enable the deployment of ITS. We need better traffic management with a holistic view of the community and all the stakeholders. For instance, let us assume a city is having a massive event, so the traffic authority decides to establish some rules for a smoother vehicle flow. Lane management application modifies the number of lanes in the same direction, changing from three north-south and three south-north to five north-south and one south-north to optimize arrival to the event, but the core of the problem (smoother vehicle flow) is not solved because most people will look for a place to park, and as the flow increases, the time to find a parking spot becomes more unmanageable. The latter is one of the reasons that calls for a systematic integration of all the applications involved. In this case, lane management and PMA could interact to assign an automated parking spot without the need for an on-site decision.

9.12.3 Diagnostic Category

This category focuses on providing diagnostic services that allow the detection of component failures that could lead to a breakdown [33] by using different types of sensors which include: (1) Powertrain sensors to check the status and functioning of the mechanical parts and engine of the vehicle in real-time, (2) sensors to monitor fuel level, (3) chemical sensors to check fluid quality, (4) temperature sensors to check the temperature of fluids or gas, (5) composition sensors to monitor engine combustion to reduce pollution, (6) chassis sensors to detect failures in the chassis systems, (7) speed and pressure sensors to monitor the pressure and speed of the wheels that allow the checking of the status of the antilock brake or the traction control system, (8) in-cabin sensors to diagnose the electric and ambience systems. This category can be improved using communication technologies to send information directly to the cloud and the service and maintenance area of the vehicle. Using a personalized vehicle registry, it is possible to identify and prevent possible car breakdowns by keeping a record and status of each vehicle part.

9.12.4 Environment Category

In the environment category information is collected from sensors deployed in or above the pavement to determine road conditions through the measurement of parameters such as road temperature, road conditions, number of chemical elements, and friction or grip of the surface (Figure 9.5). Weather prediction applications are based on surveillance, monitoring, and prediction of weather and roadway conditions to implement

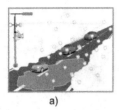 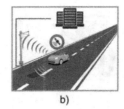 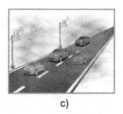

a)   b)   c)

FIGURE 9.5 ITS environment monitoring applications, (a) road weather conditions, (b) surface state, and (c) pollution management.

the appropriate management actions that improve the driving experience and mitigate the impacts of any adverse conditions [34]. Road weather applications are used to facilitate decisions on maintenance strategies and driver advisories. Weather stations and infrared sensors are deployed on roads to determine air temperatures, precipitation, as well as the presence of fog, smoke, or other conditions that could increase the risk situations for drivers or affect road maintenance decisions [35].

Road surface state applications use infrared sensors to measure the infrared radiation emitted by the surface and apply intelligent signal processing to remotely measure road parameters such as temperature, amount of water, ice, and snow [36]. One variation is the road surface anomalies monitoring application that uses sensors such as GPS, laser, infrared, and accelerometers inside vehicles to detect anomalies such as potholes or speed bumps. The collected information is used to create an anomalies map for drivers and help road managers execute infrastructure maintenance and management operations to ensure safety and comfort to drivers [37].

As mentioned, the individual application work is not successful for the creation of an ITS. Full integration and data exchange supported by cloud computing and intelligent algorithms are crucial for traffic management applications to make decisions considering not only vehicular flow but also environmental conditions and the surroundings to enable a balanced redistribution of traffic and reduce contaminants within a given zone without affecting others.

9.12.5 User Category

In the user category, sensors monitor the drivers' performance and behavior, which are essential for traffic safety and reducing accidents (Figure 9.6), using conditions such as fatigue, alcohol levels, and emotional state disorders. The technical report of the American Automobile Association (AAA) identified that drowsy driving caused 21% of deadly traffic accidents and 13% of crashes that required hospitalization [38], drunk driving caused 20% and 69% of fatally injured drivers in high-income [39] and low-income countries [40] respectively. According to the National Highway Traffic Safety Administration (NHTSA), drivers' distractions caused 10% of the fatalities in 2014 [41]. Drowsy driving warning applications are used to prevent accidents by monitoring the eyes and head motion using cameras to detect signs of drowsiness. Radar sensors monitor the car's movements and determine if the vehicle is driven erratically. Other applications can use steering angle sensors to detect an anomaly condition in the driving behavior. The driver is alerted through vibration or audio signals.

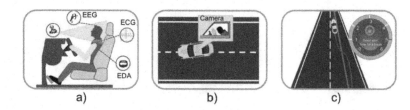

a)   b)   c)

FIGURE 9.6 ITS user monitoring applications, (a) driver's health and emotions monitoring, (b) drowsy driver warning, and (c) driver alert control.

Driver alert control applications use front-facing video cameras to track the left and right lane markings to alert the driver if the vehicle drifts outside of those lanes, thus helping reduce the probability of an accident. When the lane is not visible or erased, a camera can be used to monitor the driver by looking for signs of fatigue. The driver is alerted by a sound signal and a flashing message in the control panel of the vehicle [42]. Some authors have proposed the use of EEG sensors and artificial intelligence (AI) to detect fatigue in the driver by analyzing the EEG brain signal changes through intelligent algorithms to detect abnormal conditions [43]. Driver's health monitoring applications use thermopile, silicon photodiode, and optical and infrared sensors with LEDs for measuring vital signs of drivers such as body temperature heart, breathing rate, and blood pressure. Sensors are deployed on the steering wheel, and seats. When the application detects a problem with the driver health (e.g., a heart attack), an emergency vehicle can be called automatically.

Several works have been proposed in the literature that use ECG sensors or wireless bio sensing networks [44]. Driver's emotions recognition applications focus on detecting signs of irritation or depression that impair driving. Using Electromyogram (EMG), ECG, respiration, and Electro Dermal Activity (EDA) sensor emotions such as high or low stress, euphoria, and disappointment combined with sophisticated algorithms such as Support vector machines (SVMs) and Adaptive Neuro-Fuzzy Interference System (ANFIS), emotions can be detected and classified. In this context, several past works have focused on drivers' emotion recognition [45], evaluated emotional states in simulated environments [46], and evaluated driver's stress [47] have been published. These applications can be classified into two broad categories: first, car makers need to create non-intrusive sensors for the driver and car occupants' spots (sensors inside the seat, cameras in strategic spots) to help lessen the burden of the driving task. On the other hand, it is necessary to create intelligent algorithms, based on AI, neural networks, machine learning, computer vision, cloud computing, fog, among others to accurately identify different emotional statuses or abnormal physical conditions in the driver and the passengers to produce trustworthy notifications to the corresponding authorities or designated people.

9.12.6 Assistance Category

Pre-trip information applications collect information about different road conditions, producing several trip options for various driving routes. Parking spot locator applications allow drivers to find available parking places at locations such as streets, garages, or parking lots. Magnetometers, RFID technologies, and GPS are used to collect data from different parking spots and can offer drivers a wide variety of opportunities to park their vehicles [48]. Tourist and events applications are developed to cover the needs of travelers in unknown areas. Drivers are assisted to find the most important places in a city, empty parking slots, and routes when drivers travel to major events (sporting games or concerts). Applications use the data collected from sensors (radar sensors, cameras, inductive loops, and weather sensors) deployed near the destination place to calculate travel time and determine alternative routes according to traffic congestion or weather conditions.

On-the-fly routing information uses different sensors (such as cameras, weather sensors, radars, ultrasonic, and loops) placed on roads that collect data about traffic conditions to enable users to make informed decisions regarding alternate routes and expected arrival times. Active prediction applications anticipate the topology of the road to optimize fuel usage and assist drivers by adjusting the speed when the vehicle starts a descent or ascent.

Map download applications help drivers to get valuable information from important places or home stations. Drivers can download travel guidance maps before traveling to a new area to get directions even in a location without Internet connection. One disadvantage of this application is its high costs of installation, deployment and maintenance of the data collection, infrastructure (cameras, sensors, among others) and the communication systems to facilitate information transmission to the relevant processing centers. Many crowdsourcing applications such as Waze (http://www.waze.com), Here We Go (http://wego.here.com), and TowIt (http://towit.io) are focusing on mobile devices and the willingness of users to share information to increase the reliability and quality of the data being collected. The processing of this information will help to identify and remove false, incomplete, and redundant information so that users receive accurate information and alerts when they are distributed.

9.13 CASE STUDY SCENARIO

In this section, we present a case study that shows how sensing technologies can be integrated with information and communication technologies to improve the transportation systems and provide help and support, for example: (1) when a car is involved in a road accident due to a pothole opening suddenly and the car gets stuck inside (as shown in Figure 9.7).

1. The vehicle monitoring system detects a potentially dangerous situation using in-vehicle and outside-vehicle sensors and wearable sensors on passengers (accelerometer, which measures the vehicle's horizontal position; LIDAR which measures the distance to impact; impact sensors that detect the scale of the impact, ECG which measures the changes in a passenger heart rate, among others), however, the accident is unavoidable and the car gets stuck in a recently formed road cavity. The car immediately starts the included security and safety protocols to perform a preliminary assessment of the situation.
2. The car's central system starts a broadcast alert protocol to notify nearby drivers and pedestrians about the accident to take additional safety precautions (in example: reducing speed or taking alternate routes). (2a) At the

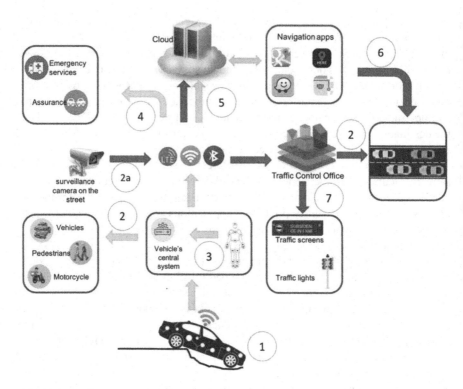

FIGURE 9.7   Case study scenario [23].

same time, using pattern recognition algorithms running in the surveillance cameras, the transport infrastructure detects the situation and activates a set of security measures for this situation such as: intelligent traffic lights can change their light management strategy, prohibiting vehicles from entering the street or blocking road access.

3. Through wearable sensors on the passenger, the car's central system receives the information and performs an assessment of passengers' health status.

4. After assessing the vehicle's damages and the passengers' health status, the central system notifies the relevant parties such as (1) the vehicle insurer, sending information such as location, insurance policy number, and preliminary damage assessment conducted from the information provided by different sensors; and (2) emergency services, sending the accident notification including, but not limited to, the number of passengers, the passenger location inside the vehicle and vital signals of each passenger, among others.

5. All the information about the accident, generated by the car's systems and protocols and the road infrastructure is sent and stored in the cloud and made available for information systems to provide further information and notifications in real-time to other drivers.

6. Location services such as Google Maps, Apple Maps, HereWe Go, and Waze can utilize the information to recalculate new or alternate routes to prevent road congestion or another accident.

7. The central system sends a notification to the transport infrastructure (traffic lights, warning screens, traffic signals) to continue sending notifications and updates regarding the accident to keep drivers and pedestrians informed about the situation.

The appraisal of ITS systems has become increasingly important in order to capture the full range of potential impacts. With the rise of such technologies as the Internet of Things and ITSs comes the inevitable question of how will they fit into our ever-changing world? The huge leaps in technological innovation we've made have thrown up more and more data about the way we live and work and increasingly this data indicates, in order for our advances to continue, the need for a shift toward more sustainable ways. Intelligent transport systems are not a new solution, however, their role in creating a secure and sustainable future for transportation will be integral. With their ability to improve and maintain traffic management through the use of connected vehicles and real-time data collection, ITSs could be applied not only for optimizing public transport services performance but also for setting the foundations for a further sustainable development.

The gradual integration of technology into roadside infrastructure has contributed to substantial savings from environmental and socio-economic perspectives [49]. With the use of environmentally damaging fossil fuels declining, ITSs will need to able to increase fuel efficiency for all connected vehicles while also being open to new, renewable technologies as and when they are brought onto the market. Increased innovation from the renewables sector, coupled with the continued and growing adoption of IT systems, will most likely lead to further development of sustainability-focused vehicle-integrated systems that provide further interconnectivity and the potential for the collection of even more actionable data. As the push for a sustainable future grows, industrial sectors that play large parts in polluting the environment should be among the first to adapt new technologies aimed at both improving their efficiency and productivity while also creating a sustainable future for continued development.

Regarding the sustainable future, The International Energy Agency's (IEA) data from 2012 suggests that CO_2 emissions from road traffic alone makes up 72% of all emissions coming from the transportation sector [50]. This statistic highlights the need for the widespread adoption of intelligent transport systems that are capable of promoting the adoption of sustainable operations and practices that will also improve efficiency and increase productivity. As with any business, the foundations you build upon will often determine how sturdy an organization you build. When it comes to the sustainable future of transport, systems designed to increase fuel efficiency, shorten journey times, optimize routes, maintain optimal performance from vehicles, and collect performance data from any connected device are exactly what transport authorities need in order to best equip them for further growth later on. But environmental protection and performance optimization aren't the only way to invest in a sustainable future, there's also the economic factor. As with any finite resource, the less fossil fuels there are, the more expensive they will be. From a transport operator's point of view, this means that every drop of fuel purchased needs to be used as efficiently as possible, wasted fuel is wasted money. While electric cars have been introduced and are slowly becoming more widespread, fossil fuel-powered vehicles are the vast majority, and transport agencies will need to be able to analyze and visualize their operational efficiency and work out where savings can be made as well as where more focus is needed. The ability to manage costs and performances of all connected vehicles could be used by operators to manage their outgoings while optimizing their performance, with the potential to use the money saved to invest in future upgrades when they're needed.

REFERENCES

1. Texas A&M Transportation Institute Technical Report 2015 Urban Mobility Scorecard, INRIX. [(accessed on 11 October 2017)]; Available online: https://static.tti.tamu.edu/tti.tamu.edu/documents/umr/archive/mobility-scorecard-2015-wappx.pdf
2. United Nations Population Fund (UNFPA) State of World Population 2011: People and Possibilities in a World of 7 Billion. United Nations Population Fund; New York, NY, USA: 2011. Technical Report. [Google Scholar]
3. Population Reference Bureau 2016 World Population Datasheet, Inform Empower Advance. [(accessed on 11 October 2017)]; 2016 Available online: https://www.prb.org/wp-content/uploads/2017/08/2017_World_Population.pdf
4. Mahmood, A., Siddiqui, S. A., Sheng, Q. Z., Zhang, W. E., Suzuki, H., & Ni, W. (2022). Trust on wheels: towards secure and resource efficient IoV networks. *Computing*, 104(6), 1337–1358. doi:10.1007/s00607-021-01040-7.
5. DIRECTIVE 2010/40/EU OF THE EUROPEAN PARLIAMENT AND OF THE COUNCIL of 7 July 2010. eur-lex.europa.eu
6. "Reducing delay due to traffic congestion. [Social Impact]. ITS. The Intelligent Transportation Systems Centre and Testbed". SIOR, Social Impact Open Repository. Archived from the original on 2017-09-05. Retrieved 2017-09-05.
7. ITSS, The ITSS Website. [Online]. Available from: https://www.ewh.ieee.org/tc/its/index.html
8. Wikipedia. Intelligent Transportation Systems. Available from: https://en.wikipedia.org/wiki/Intelligent transportation system
9. Qu, F., Wang, F., & Yang, L. (2010). Intelligent transportation spaces: vehicles, traffic, communications, and beyond. *IEEE Communications Magazine*, 48, 136–142.
10. Ezell, S.: Explaining international IT application leadership: Intelligent Transportation Systems. The Information Technology & Innovation Foundation (2010)
11. Intelligent Transportation Systems and Services
12. Contreras J., Zeadally S., Guerrero-Ibanez J.A. Internet of Vehicles: Architecture, Protocols, and Security. IEEE Internet Things J. 2017 doi: 10.1109/JIOT.2017.2690902. [CrossRef] [Google Scholar]
13. Guerrero-Ibáñez J.A., Flore-Cortés C., Zeadally S. Vehicular ad Hoc Networks (VANETs): Architecture, Protocols and Applications. In: Chilamkurti N., Chaouchi H., Zeadally S., editors. Next-Generation Wireless Technologies 4G and Beyond. 1st ed. Springer; London, UK: 2013. pp. 49–70. [Google Scholar]
14. Ezell, S. (2010). *Explaining International IT Application Leadership: Intelligent Transportation Systems*. The Information Technology & Innovation Foundation.
15. [Online] Available from: https://its.dot.gov/
16. [Online] Available from: https://www.its.dot.gov/pilots/
17. wirelessinternet/g/bldef wimax.htm
18. Wikipedia. OnStar. Available from: https://en.wikipedia.org/wiki/OnStar
19. 3rd eSafety Forum, 25 March 2004 https://citeseerx.ist.psu.edu/document?repid=rep1&type=pdf&doi=ace191a62962e5d92c2b8dcd03fbaca671e3222f

20. European Commission, Directorate-General "Information Society", Directorate C. (2004). Miniaturisation, embedded systems and societal applications. In *Unit C.5 "ICT for Transport and the Environment", "Towards Cooperative Systems for Road Transport", Transport Clustering Meeting*, USA 8 November 2004.
21. Sattar, F., Karray, F., Kamel, M., Nassar, L., & Golestan, K. (2016). Recent advances on context-awareness and data/information fusion in ITS. *International Journal of Intelligent Transportation Systems Research*, 14(1), 1–19.
22. Ben-Gal, I., Weinstock, S., Singer, G., & Bambos, N. (2016). *Frequently asked questions. Intelligent Transportation Systems Joint Program Office*. United States Department of Transportation. USA 10 November 2016.
23. Weiland, R. J., & Purser, L. B. (2000). Intelligent transportation systems. *Transportation in the New Millennium*. Transportation Research Board.
24. Guerrero-Ibáñez, J., Zeadally, S., & Contreras-Castillo, J. (2018). Sensor technologies for intelligent transportation systems. *Sensors*, 18(4), 1212.
25. Katzourakis, D. I., Lazic, N., Olsson, C., & Lidberg, M. R. (2015). Driver steering override for lane-keeping aid using computer-aided engineering. *IEEE/ASME Transactions on Mechatronics*, 20, 1543–1552.
26. Qin, Y.; Dong, M.; Zhao, F.; Langari, R.; Gu, L. (2015). Road profile classification for vehicle semi-active suspension system based on Adaptive Neuro-Fuzzy Inference System. In *Proceedings of the 2015 54th IEEE Conference on Decision and Control (CDC)*, Osaka, Japan, 14 December 2015 (pp. 1533–1538).
27. Shi, J., & Wu, J. (2017). Research on Adaptive Cruise Control based on curve radius prediction. In *Proceedings of the 2017 2nd International Conference on Image, Vision and Computing (ICIVC)*, Chengdu, China, 2–4 June 2017 (pp. 180–184).
28. Kim, S., Kim, J., Yi, K., & Jung, K. (2017). Detection and tracking of overtaking vehicle in Blind Spot area at night time. In *Proceedings of the 2017 IEEE International Conference on Consumer Electronics (ICCE)*, Las Vegas, NV, USA, 8–10 January 2017 (pp. 47–48).
29. Mehrabi, A., & Kim, K. (2015). Using a mobile vehicle for road condition surveillance by energy harvesting sensor nodes. In *Proceedings of the 2015 IEEE 40th Conference on Local Computer Networks (LCN)*, Clearwater Beach, FL, USA, 26–29 October 2015 (pp. 189–192).
30. Liu, K., Son, S. H., Lee, V. C. S., & Kapitanova, K. (2011). A token-based admission control and request scheduling in lane reservation systems. In *Proceedings of the 14th International IEEE Conference on Intelligent Transportation Systems (ITSC)*, Washington, DC, USA, 5–7 October 2011 (pp. 1489–1494).
31. Tomas-Gabarron, J., Egea-Lopez, E., & Garcia-Haro, J. (2013). Vehicular trajectory optimization for Cooperative Collision Avoidance at high speeds. *IEEE Transactions on Intelligent Transportation Systems*, 14, 1930–1941.
32. Chen, L., & Englund, C. (2016). Cooperative intersection management: a survey. *IEEE Transactions on Intelligent Transportation Systems*, 17, 570–586.
33. Nandhini, H., Sunandha, C. P., & Yamura, S. (2016). Smart parking system and slot allocation with congestion avoidance technique. *International Journal of Innovative Research in Science, Engineering and Technology*, 5. http://www.ijirset.com/upload/2016/march/105_SMART.pdf
34. Pu, L., Liu, Z., Meng, Z., Yang, X., Zhu, K., & Zhang, L. (2015). Implementing on-board diagnostic and GPS on VANET to safe the vehicle. In *Proceedings of the 2015 International Conference on Connected Vehicles and Expo (ICCVE)*, Shenzhen, China, 19–23 October 2015 (pp. 13–18).
35. Yang, J. Y., Chou, L. D., Li, Y. C., Lin, Y. H., Huang, S. M., Tseng, G., Wang, T. W., & Lu, S. P. (2010). Prediction of short-term average vehicular velocity considering weather factors in urban VANET environments. In *Proceedings of the 2010 International Conference on Machine Learning and Cybernetics*, Qingdao, China, 11–14 July 2010 (pp. 3039–3043).

36. Fedele, R., Praticò, F. G., Carotenuto, R., & Giuseppe Della Corte, F. (2017). Instrumented infrastructures for damage detection and management. In *Proceedings of the 2017 5th IEEE International Conference on Models and Technologies for Intelligent Transportation Systems (MT-ITS)*, Naples, Italy, 26–28 June 2017 (pp. 526–531).

37. Astarita, V., Vaiana, R., Iuele, T., Caruso, M. V., Vincenzo, P., & De Masi, F. (2014). Automated sensing system for monitoring of road surface quality by mobile devices. *Procedia - Social and Behavioral Sciences*, 111, 242–251.

38. Tefft, B. C. AAA Foundation for Traffic Safety. Prevalence of Motor Vehicle Crashes Involving Drowsy Drivers. Available from https://newsroom.aaa.com/wp-content/uploads/2014/11/AAAFoundation-DrowsyDriving-Nov2014.pdf (accessed on 11 October 2017).

39. World Health Organization. Global Status Report on Road Safety. Available from: https://www.who.int (accessed on 11 October 2017).

40. World Health Organization. Report on Road Traffic Injury Prevention. Available from: https://apps.who.int/iris/bitstream/10665/42871/1/9241562609.pdf (accessed on 13 January 2018).

41. James, L., & Nahl, D. (2000). Road Rage and Aggressive Driving: Steering Clear of Highway Warfare. Amherst, NY, USA, Prometheus Books.

42. Hossan, A., Kashem, F.B., Hasan, M. M., Naher, S., & Rahman, M. I. (2016). A smart system for driver's fatigue detection, remote notification and semi-automatic parking of vehicles to prevent road accidents. In *Proceedings of the 2016 International Conference on Medical Engineering, Health Informatics and Technology (MediTec)*, Dhaka, Bangladesh, 17–18 December 2016 (pp. 1–6).

43. Simon, M., Schmidt, E. A., Kincses, W. E., Fritzsche, M., Bruns, A., Aufmuth, C., Bogdan, M., Rosenstiel, W., & Schrauf, M. (2011). EEG alpha spindle measures as indicators of driver fatigue under real traffic conditions. *Clinical Neurophysiology*, 122, 1168–1178.

44. Singh, R. K., Sarkar, A., & Anoop, C. S. (2016). A health monitoring system using multiple non-contact ECG sensors for automotive drivers. In *Proceedings of the 2016 IEEE International Instrumentation and Measurement Technology Conference Proceedings*, Taipei, Taiwan, 23–26 May 2016 (pp. 1–6).

45. Reyes-Muñoz, A., Domingo, M. C., López-Trinidad, M. A., & Delgado, J. L. (2016). Integration of body sensor networks and vehicular ad-hoc networks for traffic safety. *Sensors*, 16, 107.

46. Katsis, C. D., Katertsidis, N., Ganiatsas, G., & Fotiadis, D. E. (2008). Toward emotion recognition in car-racing drivers: a biosignal processing approach. *IEEE Transactions on Systems Man and Cybernetics - Part A Systems and Humans*, 38, 502–512.

47. Deng, Y., Hsu, D. F., Wu, Z., & Chu, C. H. Feature selection and combination for stress identification using correlation and diversity. In *Proceedings of the 12th International Symposium on Pervasive Systems, Algorithms and Networks*, San Marcos, TX, USA, 13–15 December 2012 (pp. 37–43).

48. Safi, Q. K., Luo, S., Wei, C., Pan, L., Chen, Q. (2017). PIaaS: cloud-oriented secure and privacy-conscious parking information as a service using VANETs. *Computer Network*, 124, 33–45.

49. El Faouzi, N., & Klein, L. A. (2016). Data fusion for ITS: techniques and research needs. *Transportation Research Procedia*, 15, 495–512.

50. Intelligent Transportation Systems and the Sustainable Future of Transport. Available from https://www.lanner-america.com/blog/intelligent-transportation-systems-sustainable-future-transport/.

10 Advancing Resilience in Intelligent Transportation Systems

10.1 BACKGROUND

In the era of intelligent infrastructure, the convergence of digital technologies with structural control and health monitoring disciplines stands out as pivotal for strengthening resilience. The fusion of sophisticated sensors and monitoring systems with artificial intelligence (AI) and the Internet of Things (IoT) enables the real-time gathering of structural health data and information. This integration not only enhances the accuracy of structural control systems but also enables transportation networks to dynamically adjust themselves to changing environmental conditions, serving as vigilant defenders against unforeseen events [1–6].

The use of simulation and numerical analysis stands out as a powerful instrument for evaluating and enhancing structural and community resilience. By simulating scenarios that replicate dynamic environmental conditions and potential disruptions, engineers gain invaluable insights into the ways to improve the resilience of the transportation infrastructure. This approach extends beyond individual structures, offering strategic support for refining designs, validating contingency plans at a large scale, and optimizing resource allocation to systematically fortify resilience [7,8].

Predictive maintenance is also essential to a forward-thinking resilience plan since it guarantees the best possible safety and sustainability for transportation infrastructure. Drawing upon data from structural health monitoring and predictive analytics, authorities can assess current safety levels and anticipate future vulnerabilities [9–13]. By taking a proactive approach, it becomes easier to create the best repair and maintenance schedules and allocate resources efficiently to solve any problems before they become more serious. This protects both the short-term safety concerns and the long-term sustainability of transportation networks [14,15].

Teaching the next generation of engineers about resilience and sustainability concepts is a revolutionary development in strengthening structural resilience. A deep comprehension of cutting-edge technology should be combined with the incorporation of resilience and sustainability concepts to develop a new generation of civil and infrastructure engineers. With this respect, it becomes essential to take an interdisciplinary approach, which is represented in integrated courses designed to meet the unique requirements of resilience and sustainability engineering. Expertise in bridges and transportation infrastructure is taught through specialized modules that guarantee resilience engineering is integrated into all engineering schools [16,17].

DOI: 10.1201/9781032691787-10

Peering into the future, the evolution of resilient transport infrastructure demands strategic responses to emerging challenges. Anticipated trends include the integration of quantum computing and blockchain for heightened security, the establishment of international standards for resilience assessments, and a globalized approach to collaborative research and development initiatives.

The chapter underscores the importance of continuous investment in research, ongoing education, and the dissemination of best practices, emphasizing that adaptability is key, as resilience is a dynamic endeavor requiring constant innovation and vigilance.

This chapter ends with a strong appeal to flexibility, emphasizing that resilience is a journey rather than a destination that needs constant creativity and attention.

10.2 RESILIENCE AND DIGITAL TECHNOLOGIES

In an era defined by escalating uncertainties, spanning from natural disasters related to climate crisis to human-made disruptions such as wars and terrorist attacks, the imperative to strengthen the resilience of ITS becomes increasingly critical.

Resilience, a concept embraced across diverse fields such as ecosystems and communities, finds its distinct application in the context of structure and infrastructure engineering. The concept of structural resilience has been defined as "the ability of a system to reduce the chances of a shock, to absorb such a shock if it occurs, and to recover quickly after the shock" [18]. Four key components of resilience have been identified by MCEER researchers: robustness, resourcefulness, redundancy, and rapidity. Such key components have been also termed the four dimensions of resilience [18–20]. Moreover, the formulation of resilience R has been provided as the quantitative capacity to maintain a specific level of functionality for a defined amount of time (the control time T_{LC}). Accordingly, resilience R has been defined as the normalized integral of a system's functionality function Q over the T_{LC} time interval (Figure 10.1).

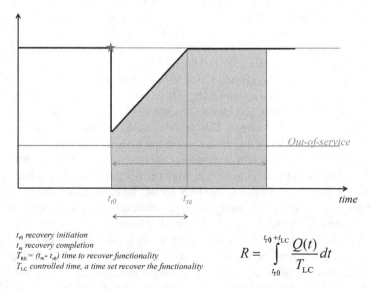

t_{r0} recovery initiation
t_{re} recovery completion
$T_{RE} = (t_{re} - t_{r0})$ time to recover functionality
T_{LC} controlled time, a time set recover the functionality

$$R = \int_{t_{r0}}^{t_{r0} + t_{LC}} \frac{Q(t)}{T_{LC}} dt$$

FIGURE 10.1 Resilience formulation.

This section draws inspiration from the larger concept of resilience and its various components to undertake a thorough analysis of the complex issues involved in strengthening the resilience of ITS. It emphasizes how crucial technology developments, especially those pertaining to digital technologies, are to achieving the resilience goals for these complex systems [16,17].

Vulnerability to hazards has increased in the complex network of interdependent modern societies, especially in the area of ITS. Transportation networks need to be carefully examined because of the numerous risks and threats they face:

- Natural disasters, accidents, or intentional attacks can lead to *physical disruptions*, impacting the seamless flow of transportation.
- A further obstacle to maintaining the demand-capacity balance is the *aging of the transportation infrastructure*, in general throughout Europe and beyond. The capacity and functionality of transportation infrastructure are directly impacted by aging and the ensuing degradation, which in turn affects resilience.
- As transportation systems become more interconnected, the risk of *cybersecurity threats* such as hacking and data breaches poses a significant challenge.
- The ever-growing demand for transportation contributes to *traffic congestion*, creating bottlenecks and increasing the susceptibility of the system to disruptions.
- Transportation systems contribute to *environmental challenges*, including pollution and resource depletion, necessitating sustainable and resilient solutions.

In facing these challenges, the integration of digital technologies becomes paramount for reinforcing the resilience of ITS:

- Real-time monitoring through sensors and advanced analysis tools enables the detection of physical disruptions promptly. Intelligent infrastructure can reroute traffic, detect-localize-evaluate structural damages, deploy emergency services, and minimize the impact of such events.
- Robust cybersecurity measures, including encryption, firewalls, and continuous monitoring, are essential. AI-driven algorithms can detect anomalous activities, providing an additional layer of protection.
- Smart traffic management systems, empowered by AI, can optimize traffic flow, alleviate congestion, and enhance the overall efficiency of transportation networks.
- Intelligent infrastructure, utilizing data analytics and machine learning, can optimize routes and modes of transportation, reducing environmental impact and promoting sustainability, life-cycle assessment (LCA), and LCA-extension-promoting measures.

Furthermore, there are ways to encourage and support the usage of recycled materials, and reuse and repurposing techniques. Sustainable goals are further aligned with the adoption of biodegradable and biocompatible materials. Resilient intelligent transportation infrastructure is further strengthened by the incorporation of, e.g., energy harvesting systems and rainwater collecting techniques, all of which fall under the general umbrella of sustainability, zero-impact initiatives, and environmental protection.

The challenges of interdisciplinary collaboration, data quality/certification, accessibility, computational complexity, and cultural, financial, and local factors are also addressed through digital technologies:

- Digital platform implementation can facilitate seamless collaboration by providing a shared and open space for different disciplines, to exchange information, knowledge, and insights.
- Digital technologies can enable the collection of high-quality, real-time data, and cloud-based solutions enhancing data accessibility for comprehensive resilience analysis.
- Advanced algorithms and computing power offered by digital technologies can overcome computational challenges, allowing for thorough and efficient resilience research.
- Digital tools can provide customizable solutions, allowing for the adaptation of resilience strategies to accommodate cultural, financial, and local distinctions.
- Blockchain technology can offer safe, immutable data records, authenticity, and control over the flow of information in ITS. Its usage promotes confidence, makes data tracing easier, and allows stakeholders to collaborate securely without sacrificing security or privacy.

Therefore, digital technologies might be crucial in promoting multidisciplinary synergies, leveraging data, and improving professional capabilities in the idea of a worldwide center for resilience control and improvement (Figure 10.2). Digital collaboration

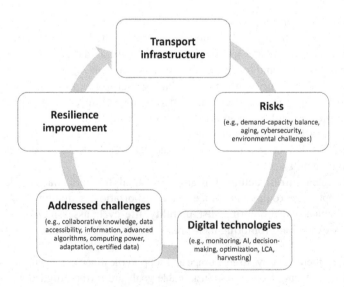

FIGURE 10.2 Digital technologies toward resilience.

platforms do, in fact, facilitate real-time communication and cooperation among multidisciplinary teams, therefore cultivating a common understanding of resilient transportation networks. A responsive and dynamic system may be built on top of such a collaborative platform by combining customization options with digital analytics and machine learning techniques. Furthermore, digital technologies combined with publicly available datasets can make model validation and calibration easier, guaranteeing the validity of resilience research. Gaining digital literacy and adaptability can be essential to utilizing new professional abilities and expanding the pool of workers qualified to handle the intricacies of resilient ITS.

In this scenario, the foundation for resilient and sustainable future transport infrastructure is established by embracing technology breakthroughs, especially digital technologies, encouraging multidisciplinary cooperation, and setting out to become a worldwide hub. This strategy demonstrates the transformative power of digital technologies in boosting resilience by ensuring the longevity of ITS against the growing challenges posed by disruptive occurrences.

10.3 CONTRIBUTION OF STRUCTURAL MONITORING TO RESILIENCE

In the dynamic landscape of intelligent transport infrastructures, this chapter aims to provide a comprehensive understanding of the integration of structural monitoring systems to contribute to resilience improvements, emphasizing key aspects such as the overview of structural monitoring, the pivotal role of sensors and systems in measuring structural state, the extraction of information on structural evolution, and the significance of real-time monitoring for early detection of unusual behaviors [16,21,22].

The base character of structural monitoring lies in the systematic and continuous assessment of a structure's condition. It involves the deployment of an array of sensors strategically placed to capture various features of structural behavior. These sensors, ranging from accelerometers and strain gauges to advanced imaging devices, collect data that can be processed to create a comprehensive picture of the structure's health. Theoretical frameworks guide the selection and placement of sensors, ensuring a holistic monitoring approach that considers different modes of structural response. In the field of transport infrastructures, this translates into a network of sensors placed to capture structural dynamics, e.g., of viaducts and bridges. Accelerometers record vibrations, strain gauges measure strain, and imaging devices provide visual data. This holistic overview enables continuous monitoring, ensuring that changes in structural behavior are captured, forming the foundation for proactive maintenance and intervention [16].

Sensors act as the eyes and ears of structural monitoring systems, capturing data that reflects the state of the structure. The theoretical approach involves selecting sensors based on the specific information needed, understanding the physics of structural response, and utilizing signal processing techniques to extract meaningful data. These systems combine data from various sensors, providing a comprehensive assessment of the structural state. With reference to bridges and viaducts, sensors play a pivotal role in measuring the structural state. Strain gauges monitor strain, acoustic emission sensors detect crack propagation, and temperature sensors gauge thermal expansion. Integrated systems process the collected data in real-time, offering an understanding of structural health. This integration of sensors and systems allows transportation authorities to collect information and, therefore, to make informed decisions regarding maintenance and intervention actions [5].

Structural evolution encapsulates the dynamic changes a structure undergoes over its service life span. In this context, data analytics and machine learning algorithms allow to extrapolation of patterns and trends within the monitored data. By identifying evolving structural behaviors, decision-makers gain insights into the deterioration progress or potential risks, forming the basis for predictive maintenance strategies. Thus, machine learning algorithms analyze historical data to predict future behavior, identifying trends that may reveal forthcoming issues. This proactive approach can allow for timely interventions, preventing structural failures and optimizing maintenance efforts [23,24].

Real-time monitoring is the cornerstone for early detection of structural anomalies. It involves the integration of sensor data with real-time processing capabilities (i.e., active monitoring). Advanced algorithms analyze incoming data, allowing for the immediate identification of deviations from expected behavior. Early detection enables swift response, minimizing the impact of structural issues on transportation networks. Therefore, for viaducts and bridges, real-time monitoring becomes a critical tool for early detection of critical conditions and to prevent hazardous conditions for users and operators, such as local collapses. It allows to analyze sensor data on the fly, triggering alerts when anomalies are detected. This ensures that transportation authorities can respond promptly to potential issues, preventing cascading failures and safeguarding the structural integrity of the transportation network.

Focusing on the quantitative contribution that a monitoring system can provide to the resilience of a transportation infrastructure, as to one of its structural components (e.g., a bridge, a tunnel), it is useful to refer to two specific phases related to monitoring for damage detection, namely diagnosis, and prognosis. Damage detection, localization, and quantification are the three foundational steps of structural health monitoring (SHM) for damage diagnosis, according to Doebling et al. [25]. The damage prognosis, or the estimate of the system's remaining life, is the last phase, or the ultimate SHM goal. The initial two phases can be performed by the prevalent vibration-based damage identification methods, without the need of adopting structural models (e.g., finite element models, or more simple analytical ones). On the contrary, damage identification, with the extension and intensity damage assessment, may be achieved by integrating vibration-based techniques with a structural model. Finally, the last phase of SHM poses a significant challenge for engineers, requiring multidisciplinary skills and predictive modeling capabilities. However, it is worth noticing how addressing this last phase could yield invaluable safety and economic advantages in structural and infrastructural management [5,11].

Infrastructure deterioration affects and reduces functionality, which is the reference function Q used to calculate infrastructure resilience [26]. Maintaining resilience requires the timely implementation of remedies to protect the infrastructure before deterioration significantly reduces its functionality. By providing vital information on the state of degradation, SHM makes it possible to restore infrastructure

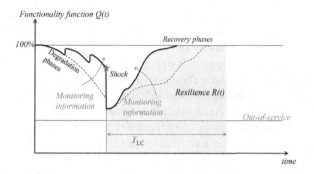

FIGURE 10.3 The value of information: monitoring contribution to resilience.

functionality and improve resilience. At several levels, resilience can be affected and improved, depending on the degree of SHM produced for the infrastructure (from damage detection to damage extension quantification). Merely being aware of the existence of damage is less important than knowing the location and the intensity of damage. As a result, a more thorough SHM procedure will have a favorable impact on resilience R by accelerating intervention and resource preparation for quicker restoration. The functionality curve and the contribution pertaining to data from a monitoring system are depicted in Figure 10.3.

Subsequent maintenance and restoration operations can be carried out, especially in the branch that was first afflicted by deterioration and eroding functionality (i.e., structural capacity and consequently operability). Knowing the location and degree of the damage, quickly allocating the required resources, accelerating the inspection and design stages, and so on, may all help to optimize interventions in the recovery branch after early deterioration and possible shocks (such as earthquakes). However, as Domaneschi and Cucuzza [11] pointed out, the data provided by the monitoring system might not have an impact on some restoration phases, including the administrative and bureaucratic ones.

10.4 CONTRIBUTION OF STRUCTURAL CONTROL TO RESILIENCE

Intelligent transport infrastructures demand adaptable control systems capable of responding to variable loading conditions. Semi-active control systems, characterized by their dynamic adjustability, play a pivotal role in optimizing structural responses in real-time [27]. This adaptability relies on the system's ability to sense changes in structural conditions factors and promptly modify its characteristics, ensuring the response is finely tuned for diverse scenarios. For example, consider a cable-supported bridge exposed to fluctuating wind speeds. Equipped with a semi-active damper system, the bridge dynamically adjusts its damping characteristics based on wind intensity. During moderate winds, the damper optimizes resistance, effectively reducing the bridge's lateral vibrations. Beyond wind scenarios, this dynamic adjustability ensures the structural integrity of the bridge under extreme loading conditions, enhancing comfort and exemplifying the practical benefits of real-time responsive and adaptation.

The integration of semi-active or active dampers can introduce redundancy and robustness into the structural system [28], as resilience components. Redundancy protects against local failures or outages, ensuring the overall functionality of the system even if one damper unit experiences issues. Simultaneously, robustness is evident in the system's ability to adapt to uncertainties, maintaining a satisfactory level of structural performance and stability. In a scenario involving a long-span cable-supported bridge equipped with an active or semi-active damping system, the importance of redundancy becomes evident. In the event of a local failure in one damper unit, the redundancy in the control system allows adjacent dampers to seamlessly compensate, preventing a complete system failure. Such an event is evaluated by Domaneschi et al. [28] by performing numerical simulations that reproduce such a compensating scenario, highlighting the benefits of such compensation in terms of resilience improvement. This underscores the significance of adaptive control solutions in maintaining operational continuity and highlights the efficiency of the control strategy when confronted with unexpected component failures.

Thus, the real-time modification of damper characteristics termed *adaptation* enhances the rapidity dimension of resilience, enabling the control system to respond swiftly to changing conditions. This capability minimizes downtime, optimizes operational efficiency, and ensures the structure's immediate adaptation to unforeseen events. Such characteristic has been also termed by Domaneschi et al. [28] as *Immediate Resilience*. For example, consider a highway overpass subjected to sudden and intense seismic activity. Equipped with an active damper system featuring real-time adjustment capabilities, the structure senses seismic forces and instantly adapts its damping properties. This automatic and almost immediate reaction significantly reduces the structure's response to the earthquake, reasonably minimizing damage and downtime. Since the control system guarantees that modifications are made rapidly and effectively, the structure may quickly return to regular operation following a seismic event, establishing the rapidity dimension of resilience.

The theoretical understanding of adaptive responses, resilience dimensions, and the practical examples presented collectively emphasize the multifaceted benefits of structural control in intelligent transport infrastructures. Indeed, by strategically implementing control strategies, these infrastructures not only mitigate the impact of external disturbances but also ensure continued functionality and durability in the face of diverse challenges. This holistic approach safeguards against specific threats and fosters a resilient and adaptable transportation infrastructure ready to meet the demands of an evolving environment.

When considering the quantitative impact that a structural control system can have on a transportation infrastructure's resilience, it is useful to highlight the contribution that the implementation of such systems can offer to the functionality curve. Observing Figure 10.4, the effect of a structural control system on the initial degradation branch, where it can influence the structure's functionality by gradually reducing it, is evident. This condition can be represented in real scenarios, for instance, by fatigue damage accumulation in metallic elements due to load cyclicality, such as wind or traffic effects. In this context, literature has demonstrated how appropriate adaptive control devices can mitigate the effects of cyclic loads [29,30].

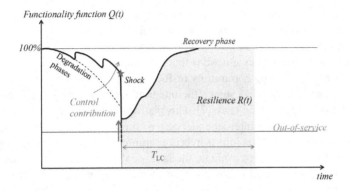

FIGURE 10.4 The structural control contribution to resilience.

Progressing along the functionality curve in Figure 10.4 until a hypothetical occurrence of an extreme event like a severe earthquake, the control system can effectively mitigate the shock's effects, significantly reducing the structural response in terms of internal actions and deformations. Thus, the so-called *controlled* structure prevents a far higher state of damage compared to what would occur if such a control system was not implemented (the *uncontrolled* structure) [31,32].

10.5 MODELING TOWARD ASSESSMENT AND IMPROVEMENT OF RESILIENCE

In the broad field of ITS, resilience stands as a crucial objective and a significant challenge. It demands a sophisticated interaction between comprehension and proactive measures, with modeling serving as a critical tool in managing and understanding such a complex environment.

Establishing robust and operational transportation networks with respect to standard hazard conditions represents just one aspect of the effort. Safeguarding these systems against unexpected disaster disruptions as the result of new loading related to climate crises or multi-hazard conditions is equally imperative. Therefore, the imperative is not only to see to past conditions but also to the future ones, to think and design infrastructures for standard, realistic, and unforeseen scenarios.

This chapter presents an overview of the many and varied aspects of modeling, emphasizing both its fundamental importance in resilience evaluation and its crucial role in coordinating development plans. With this aim, it is of paramount importance to recognize how the analysis of the specific transportation infrastructure cannot disregard the interdependencies. Indeed, examining the specific network and links between other infrastructures provides important information about how vulnerable transportation infrastructures vulnerable to hazards are and how exposed they are to them. Gaining an understanding of these interactions can be essential to improve resilience.

Given this requirement and the growing frequency of catastrophes the task of developing large-scale simulation models has taken on significant meaning. Recently, a number of simulation techniques have been created to investigate how communities react to natural catastrophes [7]. The primary goal of these models is to assist decision-makers in emergency operations by enabling them to recognize the implications of the catastrophe and build a comprehensive picture of the situation. They may also be used to create strategic actions that will improve preparedness and resource planning, hence increasing community resilience.

Recently, the literature has shown the integration of all computing resources into a single platform by Marasco et al. [7]. This platform can replicate the specific infrastructure response and its interdependencies with other networks. It incorporates a real-time simulation feature into a hybrid community concept. Using appropriate models, the platform aims to evaluate the large-scale seismic resilience and vulnerability of critical infrastructures while accounting for their interdependencies.

The platform incorporates multiple layers, including distribution networks, buildings, and socio-technical networks, in addition to the transportation network. Through particular models, interdependencies between the various levels have also been devised and put into practice. In addition to first aid procedures in post-disaster scenarios, the platform enables the consideration of the emergency evacuation process using an agent-based model at a large scale [2,3].

The platform may simulate bridge closures in response to external dangers by concentrating on the transport infrastructure network. This allows for the creation of scenarios that can be simulated, allowing for the planning of different routes. In this way, the platform enables the examination of the transport infrastructure network's resilience as well as the potential use of substitute measures that may enhance the circumstances following an event. One example of this interaction between the built environment and the transport network is also simulated. First, debris created by, say, local damage, can impact the network's efficiency state. The platform can take these factors into account and estimate the time required for recovery as well as the steps required to get over these obstacles [7,33].

In addition to evaluating the debris extension, these methods enable the investigation of a community's reaction to a disruptive event and the development of resilience plans to reduce performance losses and recovery times.

10.6   THE ROLE OF EDUCATION IN BUILDING RESILIENCE

The development of resilience is a key objective in the dynamic field of ITS. Though scientific and technical progress undoubtedly opens the door to future-proof solutions, training experts specialized in transportation infrastructure management, design, and decommissioning is essential to unifying our transportation system. It becomes essential for the creation of strong systems to train the upcoming generation of structural and civil engineers in the concepts of sustainability and resilience, as well as in cutting-edge technology.

In academic engineering education, resilience and sustainability should be established as core values in order to foster long-lasting change. To address this, multidisciplinary and integrated courses tailored to the needs of the upcoming generation of resilience and sustainability engineers must be developed.

In order to integrate many subjects including data analysis, digital infrastructure, environmental sciences, urban planning, and civil engineering, these educational pathways need to go beyond traditional disciplinary boundaries. Future engineers will be better prepared to understand the complexities of resilient design and to conceive environmentally aware solutions with the help of such a comprehensive approach.

A paradigm change is apparent when considering bridges and transportation infrastructures: more specific knowledge on resilience is required. The dynamic character of disruptive forces demands a more thorough comprehension of how these systems tolerate shocks and bounce back from them. Consequently, there is a pressing need to incorporate specialized courses that examine case studies, examine failure modes, and present fortification strategies against a variety of adversities into educational modules and programs. These include those mandated by regulations, those resulting from contemporary challenges like the climate crisis, and even those that are yet to come but that can (and should) be anticipated in their complexity.

Even though it is still in its early stages, some universities and technological institutes already provide resilience engineering for infrastructures. The real effect, nevertheless, will come from its widespread acceptance by all universities that offer engineering degrees. The integration and growth of resilience engineering in academic environments offer a chance to train a new generation of engineers with the knowledge and skills required to strengthen our transportation infrastructure, and communities, against unforeseen challenges.

Thus, the development of engineering education goes beyond traditional bounds. It is a call for a radical change, an evolution that inspires a mindset in addition to technical understanding. This resilient and sustainable way of thinking lays the groundwork for future engineers to create solutions that go beyond accepted boundaries.

10.7   FUTURE TRENDS AND RECOMMENDATIONS

The trajectory of ITS is driven by a continuous flow of innovation and adaptation actions. Looking ahead, a number of significant patterns are revealed, which indicate a future that will require awareness and careful planning. These patterns, along with developing technology and evolving social demands, may open the door for revolutionary innovations that could shape the resilient transportation systems of the future.

The future of ITS can be driven by the ubiquity of AI and Machine Learning (ML). They can be essential for strengthening resilience because of their capacity to handle enormous amounts of data, optimize traffic patterns, predict system behavior, and identify vulnerabilities. Allocating funds to research and development would be recommended to fully leverage the potential of AI and ML for anomaly detection, real-time system adaptation, and predictive modeling within transportation networks.

Connected with AI and ML is the introduction of autonomous vehicles, announcing a paradigm shift in transportation. The seamless integration of these vehicles within existing infrastructure may meet challenges, e.g., with respect to compatibility. Starting to think about new infrastructure or innovating existing infrastructure to be able to integrate the latest technologies, they also seek to anticipate future developments in technology, becoming a cornerstone for accommodating and optimizing the integration of autonomous vehicles, and improving the safety, efficiency, and overall resilience of transportation networks.

Sustainability emerges as a non-negotiable aspect in the evolution of transportation systems, adopting solutions that not only strengthen resilience but also harmonize with the environment. It is recommended that sustainable practices be included in the transportation infrastructure by supporting environmentally friendly designs, the integration of renewable energy sources, and techniques for mitigating negative environmental impacts.

As systems become increasingly interconnected, the vulnerability to cyber threats amplifies. Safeguarding transportation networks against cyber-attacks becomes of paramount importance. The recommendation would be to strengthen against possible weaknesses and attacks by investing in strong cybersecurity measures, putting encryption procedures into place, and fostering a culture of data protection.

Resilience must be the primary focus of planning and design, not just a secondary consideration. It is recommended that resilience be included in policy frameworks, urban planning, and engineering education as a core design concept. This will encourage a proactive approach to strengthening systems against disruptions and facilitating adaptability.

Regulations need to change in order to keep up with society's demands, hazards progress, and technological breakthroughs. It would be recommended to encourage flexible rules that ensure safety in transportation networks and promote sustainability and innovation. Stakeholders should be involved in cooperative efforts to adopt rules in response to the evolving situation.

Communities are stronger when they're allowed an input in how transportation networks are designed. In order to ensure that systems meet the needs and expectations of the people they serve, it is advised that community participation be promoted through outreach programs, educational efforts, and participatory planning procedures.

The problems that transportation networks face are worldwide in magnitude and go beyond national borders. Promoting a global effort toward resilient transportation networks, international cooperation and knowledge-sharing platforms for the exchange of best practices, lessons learned, and creative solutions would be advisable.

ITS will evolve and adapt throughout time rather than reaching a single, static state. Developing ethical frameworks, embracing emerging technology, creating flexible regulations, and encouraging interdisciplinary cooperation are all efforts that aim toward strengthening resilience.

Future directions call for us to cultivate an environment in which technological advancements prioritize ethical considerations, where policy frameworks are flexible but inclusive, where creativity is moderated with responsibility, and where cooperative efforts lay the groundwork for resilient transportation systems that meet the needs of diverse communities.

REFERENCES

1. Domaneschi, M. (2020). Resilience of bridges in infrastructural networks. In Z. Wu, X. Lu, & M. Noori (Eds.), Resilience of Critical Infrastructure Systems: Emerging Developments and Future Challenges, 1st Edition, CRC Press. https://doi.org/10.1201/9780367477394, Boca Raton FL, 177–188.

2. Battegazzorre, E., Bottino, A., Domaneschi, M., & Cimellaro, G. P. (2021), IdealCity: a hybrid approach to seismic evacuation modeling. *Advances in Engineering Software*, 153, 102956.

3. De Iuliis, M., Battegazzorre, E., Domaneschi, M., Cimellaro, G. P., & Bottino, A. G. (2023). Large scale simulation of pedestrian seismic evacuation including panic behavior. *Sustainable Cities and Society*, 94, 104527.

4. Martinelli, L., Domaneschi, M., Cucuzza, R., & Noori, M. (2023). Service-life extension of transport infrastructure through structural control. In *Eighth International Symposium on Life-Cycle Civil Engineering IALCCE 2023*, Milan, Italy.

5. Noori, M., Rainieri, C., Domaneschi, M., & Sarhosis, V. (2023). *Data Driven Methods for Civil Structural Health Monitoring and Resilience: Latest Developments and Applications* (pp. 1–341). CRC Press. DOI: 10.1201/9781003306924.

6. Nagarajan, S. M., Devarajan, G. G., Ramana, T. V., Bashir, A. K., & Al-Otaibi, Y. D. (2024). Adversarial deep learning based Dampster-Shafer data fusion model for intelligent transportation system. *Information Fusion*, 102, 102050.

7. Marasco, S., Cardoni, A., Noori, A. Z., Domaneschi, M., & Cimellaro, G. P. (2021), Integrated platform to assess seismic resilience at the community level. *Sustainable Cities and Society*, 64, 102506.

8. Mitoulis, S. A., Domaneschi, M., Cimellaro, G. P., & Casas, J. R. (2022). Bridge and transport network resilience - a perspective. *Proceedings of the Institution of Civil Engineers - Bridge Engineering*, 175(3), 138–149. DOI: 10.1680/jbren.21.00055.

9. Binder, M., Mezhuyev, V., & Tschandl, V. (2023). Predictive maintenance for railway domain: a systematic literature review. *IEEE Engineering Management Review*. DOI: 10.1109/EMR.2023.3262282.

10. Domaneschi, M., Cimellaro, G. P., Xie, L., Bruneau, M., Wu, Z., Didier, M., Noori, M., Mufti, A., Lu, X., Lu, X., Ou, J., Sheikh, S., Zhou, Y., Yoda, T., Taciroglu, E., Häring, I., Sextos, A. (2021). Present and future resilience research driven by science and technology. *International Journal of Sustainable Materials and Structural Systems*, 5(1/2), 50–89. https://www.eng.buffalo.edu/~bruneau/IJSMSS%202020%20Resilience.pdf

11. Domaneschi, M., & Cucuzza, R. (2023). Structural resilience through structural health monitoring: a critical review. In *Data Driven Methods for Civil Structural Health Monitoring and Resilience: Latest Developments and Applications* (pp. 1–13). CRC Press. DOI: 10.1201/9781003306924-1.

12. Domaneschi, M., Martinelli, L., Cucuzza, R., & Noori, M. (2023). Service-life extension of transport infrastructure through structural health monitoring. *Eighth International Symposium on Life-Cycle Civil Engineering IALCCE 2023*, Milan, Italy.

13. Kokane, C. D., Mohadikar, G., Khapekar, S., Jadhao, B., Waykole, T., & Deotare, V. V. (2023). Machine learning approach for intelligent transport system in IOV-based vehicular network traffic for smart cities. *International Journal of Intelligent Systems and Applications in Engineering*, 11(11s), 06–16.

14. Chaudhary, R., Jindal, A., Aujla, G. S., Aggarwal, S., Kumar, N., & Choo, K.-K. R. (2019). BEST: blockchain-based secure energy trading in SDN-enabled intelligent transportation system. *Computers and Security*, 85, 288–299.

15. Ganin, A. A., Mersky, A. C., Jin, A. S., Kitsak, M., Keisler, J. M., & Linkov, I. (2019). Resilience in Intelligent Transportation Systems (ITS). *Transportation Research Part C: Emerging Technologies*, 100, 318–329.

16. Morgese, M., Domaneschi, M., Ansari, F., Cimellaro, G. P., & Inaudi, D. (2021). Improving distributed FOS measures by DIC: a two stages SHM. *ACI Structural Journal*, 18(6), 91–102.

17. Mitoulis, S. A., Domaneschi, M., Casas, J. R., Cimellaro, G. P., Catbas, N., Stojadinovic, B., & Frangopol, D. M. (2022). Editorial. The crux in bridge and transport network resilience - advancements and future-proof solutions. *Proceedings of the Institution of Civil Engineers - Bridge Engineering*, 175(3), 133–137. DOI: 10.1680/jbren.2022.175.3.133.

18. Bruneau, M., Chang, S. E., Eguchi, R. T., Lee, G. C., O'Rourke, T. D., Reinhorn, A. M., Shinozuka, M., Tierney, K., Wallace, W. A., & Von Winterfeldt, D. (2003). A framework to quantitatively assess and enhance the seismic resilience of communities. *Earthquake Spectra*, 19(4), 733–752.

19. Bruneau, M., & Reinhorn, A. M. (2006). Overview of the resilience concept. *Proc., 8th U.S. National Conf. on Earthquake Engineering, Earthquake Engineering Research Institute (EERI)*, Oakland.

20. Bruneau, M., & Reinhorn, A. M. (2007), Exploring the concept of seismic resilience for acute care facilities. *Earthquake Spectra*, 23(1), 41–62.

21. Zakutynskyi, I. R. (2023). IoT system architecture for monitoring and analyzing public transport data. *Multidisciplinary Science Journal*, 5, 1–8, 2023ss0103.

22. Guo, A. J., Lin, J., & Li, X. (2020). Data mining algorithms for bridge health monitoring: Kohonen clustering and LSTM prediction approaches. *The Journal of Supercomputing*, 76, 932–947.

23. Ye, X. W., Jin, T., & Yun, C. B. (2019). A review on deep learning-based structural health monitoring of civil infrastructures. *Smart Materials and Structures*, 24(5), 567–585.

24. Sun, L., Shang, Z., Xia, Y., Bhowmick, S., & Nagarajaiah, S. (2020). Review of bridge structural health monitoring aided by big data and artificial intelligence: From condition assessment to damage detection. *Journal of Structural Engineering*, 146(5), 04020073.

25. Doebling, S., Farrar, C., Prime, M., & Shevitz, D. (1996). Damage identification and health monitoring of structural and mechanical systems from changes in their vibration characteristics: a literature review. Los Alamos National Lab., Los Alamos, NM.

26. Vishwanath, B., & Banerjee, S. (2019). Life-cycle resilience of aging bridges under earthquakes. *Journal of Bridge Engineering*, 24, 04019106.

27. Domaneschi, M. (2012). Simulation of controlled hysteresis by the semi-active Bouc-Wen model. *Computers and Structures*, 106–107, 245–257.

28. Domaneschi, M., Martinelli, L., Perotti, F. (2016). Wind and earthquake protection of cable-supported bridges. *Proceedings of the Institution of Civil Engineers - Bridge Engineering*, 169(3), 157–171.

29. Martinelli, L., & Domaneschi, M. (2017), Effect of structural control on wind fatigue mitigation in suspension bridges. *International Journal of Structural Engineering*, 8(4), 289–307.

30. Pourzeynali, S., & Datta, T. K. (2005. Reliability analysis of suspension bridges against fatigue failure from the gusting of wind. *Journal of Bridge Engineering*, 10(3), 262–271.

31. Domaneschi, M., & Martinelli, L. (2014). Extending the benchmark cable-stayed bridge for transverse response under seismic loading. *Journal of Bridge Engineering ASCE*, 19(3), 4013003.

32. Domaneschi, M., & Martinelli, L. (2016). Earthquake resilience-based control solutions for the extended benchmark cable-stayed bridge. *Journal of Structural Engineering, ASCE*, 142(8), 4015009.

33. Domaneschi, M., Cimellaro, G. P., & Scutiero, G. (2019). A simplified method to assess generation of seismic debris for masonry structures. *Engineering Structures*, 186, 306–320.

Index

Note: **Bold** page numbers refer to tables and *italic* page numbers refer to figures.

Printed in the United States
by Baker & Taylor Publisher Services